Indigenous Responses to Mining in Post-Conflict Colombia

This book examines Indigenous responses to mining and their connection to peacebuilding, focusing on the experience of the Nasa Indigenous people of North Cauca during the most recent Colombian post-agreement transition.

Amid an armed conflict that has disproportionally affected and targeted the Nasa, as well as ongoing processes of dispossession and oppression, the Nasa have built a tradition of organised, peaceful resistance. This book examines the nature of their responses to mining and how this is linked to peacebuilding, with a focus on how resistance is shaped and enacted to respond to the relationship mineral extraction has with violence and peace. The work is exploratory, ethnographic and interdisciplinary in nature, sitting in the intersection between the anthropology of mining, development studies and peace and conflict studies. The author presents and analyses narratives, participant responses, and her own experiences to illustrate the context and interconnected processes shaping Nasa responses to mining during this transition period. The book will bring international readers closer to these intricate dynamics, where access is otherwise limited because of security, cultural, linguistic and other barriers. The book provides a novel perspective on post-conflict mining governance by focusing on the Nasa's active role in responding to mining in a post-agreement, transitional context. It highlights, and encourages engagement with, the often-overlooked role of morality in debates about nature and development.

This book will be of great interest to students, scholars and practitioners of the extractive industries, natural resource management, conflict management and peacebuilding, Indigenous Peoples and Latin American studies.

Diana Carolina Arbeláez Ruiz is Postdoctoral Researcher at the University of Eastern Finland's Faculty of Social and Business Studies and Industry Fellow at the Sustainable Minerals Institute, The University of Queensland, Australia. She holds a PhD from The University of Queensland and has worked as a researcher and consultant examining the interaction between mining and communities in social and armed conflict, and post-conflict environments.

Routledge Studies of the Extractive Industries and Sustainable Development

The Anthropology of Resource Extraction
Edited by Lorenzo D'Angelo and Robert Jan Pijpers

Andean States and the Resource Curse
Institutional Change in Extractive Economies
Edited by Gerardo Damonte and Bettina Schorr

Stakeholders, Sustainable Development Policies and the Coal Mining Industry
Perspectives from Europe and the Commonwealth of Independent States
Izabela Jonek-Kowalska, Radosław Wolniak, Oksana A. Marinina and Tatyana V. Ponomarenko

The Social Impacts of Mine Closure in South Africa
Housing Policy and Place Attachment
Lochner Marais

Local Communities and the Mining Industry
Economic Potential and Social and Environmental Responsibilities
Edited by Nicolas D. Brunet and Sheri Longboat

The Shaping of Greenland's Resource Spaces
Environment, Territory, Geo-Security
Mark Nuttall

Indigenous Responses to Mining in Post-Conflict Colombia
Violence, Repression and Peaceful Resistance
Diana Carolina Arbeláez Ruiz

For more information about this series, please visit: www.routledge.com/Routledge-Studies-of-the-Extractive-Industries-and-Sustainable-Development/book-series/REISD

Indigenous Responses to Mining in Post-Conflict Colombia

Violence, Repression and Peaceful Resistance

Diana Carolina Arbeláez Ruiz

Routledge
Taylor & Francis Group
LONDON AND NEW YORK

earthscan
from Routledge

First published 2024
by Routledge
4 Park Square, Milton Park, Abingdon, Oxon OX14 4RN

and by Routledge
605 Third Avenue, New York, NY 10158

Routledge is an imprint of the Taylor & Francis Group, an informa business

British Library Cataloguing-in-Publication Data
A catalogue record for this book is available from the British Library

ISBN: 978-1-032-12926-6 (hbk)
ISBN: 978-1-032-12929-7 (pbk)
ISBN: 978-1-003-22689-5 (ebk)

DOI: 10.4324/9781003226895

Typeset in Times New Roman
by Taylor & Francis Books

To Martin, Julieta and Juanes.

To those who died, and those who live, in the struggle for Colombian peace.

Contents

Illustrations

Maps

Tables

Acknowledgements

The list of people whom I would like to thank is long, as was the journey to bring this book to fruition. This is an adaptation of my doctoral dissertation, a process that involved four years of collaboration with Indigenous organisations in Colombia with periods of study, analysis and writing at several research institutions both during the doctorate and beyond. First, I would like to thank the Association of Indigenous Authorities of North Cauca (ACIN) and the Indigenous Reserve of Canoas for their participation in this research. I thank the Thuthenas of ACIN and the Authorities of the Canoas Cabildo, and each of the Nasa community members that I was fortunate to meet during the fieldwork. I value their openness and insight and admire their courage and their hard work amid the upsurge of violence against their leaders and communities. I am extremely grateful for the friendship and collaboration of Marco Fidel Mosquera, Nelson Valencia and Rosalba Velasco, and to the Valencia Velasco family for hosting me while in Canoas. I appreciate the valuable support of Rodolfo Pilcue and Katherine Martinez Muelas. In gratitude, I pay my respects to the memory of Albeiro Camayo, who was Coordinator of ACIN's Guardia Indígena during my stay in North Cauca, and helped me plan and coordinate safe access to several Indigenous Reserves. Like many other Nasa leaders, Albeiro was murdered while he worked to protect the territories and rights of his community. I am grateful for the openness and generosity of 60 interview participants from ACIN, the Canoas Indigenous Reserve, the broader Nasa community, and various Colombian and international organisations.

In conducting this research, I sought the free prior and informed consent of the Canoas Cabildo and ACIN Thutenas or Councillors. In addition to presenting findings and receiving feedback during the research, the commitments I made included not to publish the research until these organisations had the opportunity to provide feedback on the findings. I received this feedback in 2019. I appreciate the time members of the participating Nasa organisation invested in attending my presentations and reviewing the report I offered.

I am grateful to my thesis advisors Martha Macintyre, Deanna Kemp and Chris Anderson. Martha Macintyre was an exceptional thesis advisor. I am forever indebted to her and honoured by the privilege of her guidance and friendship in this journey. I am at a loss of words to thank Martha. Her generous, focused and clear advice made a world of difference to me. Martha is a wonderful mentor of women and of women of non-English speaking backgrounds. I cannot but marvel at the scale of Martha's contribution to education and research, her patience and her unwavering commitment. I am profoundly grateful to Deanna Kemp for her generous support, trust, encouragement and careful feedback. Chris Anderson was a source of support and encouragement, and I gratefully appreciate his generosity towards me.

I thank the people in the Sustainable Minerals Institute and University of Queensland community who have helped and accompanied me along the process. John Owen, Nick Bainton and Serge Loode offered valuable comments on dissertation drafts. Serge Loode generously shared his knowledge of conflict and peace studies concepts with me. Neil MacIntyre, David Brereton and Daniel Franks offered generous support for the research or book writing. Elaine Wightman, Michelle Rowland, Melissa Miller, Tess Dobinson and Bronwyn Battersby helped me navigate the procedures and administration. Mali Moazen helped me with IT questions.

I thank Erin Fitz-Henry, from the University of Melbourne, and Susan Hemer, from the University of Adelaide, for their thoughtful and helpful feedback on dissertation drafts; and three anonymous reviewers who provided invaluable recommendations based on the book proposal. Hanna Lehtimäki, from the University of Eastern Finland, offered helpful comments on the introduction chapter. Jenni Lewington, from the Editing Edge, provided copy-editing services for the original dissertation manuscript.

During the data analysis and write-up, I had the fortune of spending a term at the Peace Research Institute Oslo, where I felt extremely welcomed and supported. I am grateful to Kristian Berg Harpviken, Kristian Hoelscher and the administration team led by Lene Kristin Borg for opening the doors to me. I give thanks to Catherine Bye, Wenche Hauge, Ebba Tellander, Elizabeth Rosvold, Noor Jdid and Anne Lene Hompland for their friendship and company.

I am grateful to Rauno Sairinen at the University of Eastern Finland for the freedom he has enabled which allowed me to work on and complete this book adaptation. My gratitude goes to Hanna Lehtimäki for welcoming me in the UEF Kuopio campus. UEF has been a supportive, peaceful and nurturing environment for my research and writing pursuits.

I thank the team at Routledge, in particular Editor Hanna Ferguson, for their patience and support.

Throughout what was a demanding research project, it was important to make sure that the mind and the body were in the right place. I am grateful to John Morton, Miriam Latif Sandbaek, John Uhr-Delia, Ashish Misra and Astor Villota.

I thank my family and friends in Australia for their hospitality and support during my studies, in particular my brother, Juan, and his family; my mother-in-law, Patricia Griffith; and the Flynn Smart family.

I would not have been able to complete this manuscript without the unconditional love, care and support of my husband, Martin, and my mother, Julieta, who have always believed in me. Martin was with me all along, in the distance and in the overwhelming proximity of my questions about prepositions, synonyms and polite expression. This is the third large manuscript of mine that Martin proofreads. He is a committed husband. My mother was a constant presence all along, while in Colombia, Brisbane and Oslo. Julieta looked after me so that I could concentrate on the research. She proofread and gave me feedback on the Spanish language documents that were part of the dissertation process, including a Spanish language summary report. I thank Martin and Julieta profoundly and forever. They are the best travel partners I could ever aspire to have. I am delighted that they are both coming along to the final part of this writing journey, in Finland.

Abbreviations

ACIN	Association of Indigenous Authorities of North Cauca (Asociación de Cabildos Indígenas del Norte del Cauca)
ACONC	Association of Afro-descendant Councils of North Cauca (Asociación de Consejos Comunitarios del Norte del Cauca)
ANUC	National Association of Campesino Users (Asociación Nacional de Usuarios Campesinos)
CNOA	First Congress of the Black Peoples of Colombia
CREER	Regional Centre on Responsible Business and Enterprise (Centro Regional de Empresas y Emprendimientos Responsables)
CRIC	Cauca Regional Indigenous Council
DANE	National Department of Statistics (Departamento Nacional de Estadísticas)
EIS	Environmental Impact Study
ELN	National Liberation Army (Ejercito de Liberación Nacional)
EPL	Popular Liberation Army (Ejercito Popular de Liberación)
ESMAD	Mobile anti-riot squadron
FARC	Revolutionary Armed Forces of Colombia – Army of the People
FENSOAGRO	Fensoagro
Foro	National Forum for Colombia (Foro Nacional por Colombia)
FPIC	Free, prior and informed consent
ICMM	International Council on Mining and Metals
IIED	International Institute for Environment and Development
ILO	International Labour Organization
ILO 169	International Labour Organisation Convention on Indigenous and Tribal Peoples, Convention No 169 (1989)
M-19	19th April Revolutionary Movement (Movimiento Revolucionario 19 de abril)
MAQL	Quintín Lame Armed Movement (Movimiento Armado Quintín Lame)
NRGI	Natural Resource Governance Institute
OECD	Organization for Economic Cooperation and Development

ONIC	National Indigenous Organisation of Colombia (Organización Nacional Indígena de Colombia)
RUCOM	Unique Registry of Mineral Vendors (Registro Único de Comercializadores de Minerales)
UN	United Nations
UNDRIP	United Nations Declaration on the Rights of Indigenous People (2007)
UNHCHR	United Nations High Commission for Human Rights
WBCSD	World Business Council for Sustainable Development

1 Introduction

One evening in early 2017, I was staying at the Nasa Indigenous Reserve of Canoas in the mountains of Cauca, Colombia, when I heard an enthusiastic community radio announcement: "*The Authorities of the Canoas Indigenous Reserve invite all community members to join tomorrow's working bee against mining! To defend life, Mother Earth, our communities and our territories! Bring your lunch, machetes, shovels and lots of spiritual strength!*" The morning after, my Nasa host and I joined members of the Nasa Indigenous community at the meeting point in the Canoas Indigenous Reserve. That was one of many days I spent learning about how Nasa communities and organisations responded to mining, amidst Colombia's peace agreement negotiation and implementation with the guerrillas of The Revolutionary Armed Forces of Colombia (FARC).[1] This was the subject of my exploration from 2015 to 2018 with the collaboration of Nasa organisations from North Cauca.

I sought to understand community responses to mining and their moral underpinning. How people judge whether and under what conditions an economic activity, such as mining, is acceptable or desirable, shapes community positions and actions towards that activity. In this context, we need to turn to the role of the 'moral economy'. The moral economy is an understanding of the economy that concentrates on morality, rather than atomised agency, as the necessary context to study the economy (Owen, 2009). The notion of the moral economy gains particular relevance in societies undergoing conflict-to-peace transitions where some of the divisions have moral undertones. I contend that at the core of Nasa resistance to mining are fundamental community moral principles; that for many Nasa, mining is a continuation of armed conflict, violence, oppression and colonisation processes that started with the Spanish invasion in 1492; that in consequence, many Nasa resist mining; and that the resistance is neither monolithic nor without the complexities that accompany material poverty. In all, I shall argue that anyone aiming to contribute to dialogue, peace or development initiatives related to mining, at this sensitive time in Colombia, needs to engage with the moral debates about mining and with the specificity of community responses to mining, in order to avoid fuelling further conflict.

But to return to the Nasa of North Cauca: Who are the Nasa? Where did this working bee against mining happen? Later, I shall go into some detail

DOI: 10.4324/9781003226895-1

about the Nasa and North Cauca to set this study in context. For now, a few brief points. If you are from Colombia, you are likely to have heard of the Nasa. They are one of the most organised Indigenous peoples of Colombia, part of a well-established resistance movement, active in advancing the Indigenous rights agenda in their territory and more broadly in Colombia (Mayr Maldonado & Olmedo Martínez, 2016). They have contributed to this through proposals, dialogue and large-scale social mobilisation. Like other Indigenous Colombians, the Nasa are among the poorest and most affected by the armed conflict and by illegal mining. North Cauca covers around 13 municipalities in the northern area of Colombia's Cauca department. It encompasses the foothills of the Western Andean chain, inter-Andean valleys and rivers that supply water to human settlements and agriculture from large cities and industrial agriculture to small towns and subsistence agriculture. Within that territory, the approximately 112,000 Nasa have 17 Indigenous Reserves (*Resguardos*) and 20 Indigenous Authorities (*Cabildos*). Colombia's 1991 Constitution grants these communities administrative, legal, judiciary, educational and health autonomy. Opposition to mining, while not the only position taken, was prevalent among the Nasa participants of this study and was the official stance of the Indigenous Authorities of many *Resguardos*, as is clear from the community radio invitation.

Let us attend to that community radio voice. A loud speaker on a motorcycle played the carefully voiced over, musicalised and compelling call to action. The background music was packed with images of resistance, delivered in the chorus of Mercedes Sosa's song *Hermano Dame tu Mano* (Brother Give me your Hand). As the Argentinian social protest singer's voice faded in the distance, the song spoke about how the land awaits people's action. The next morning many residents of the Canoas Indigenous Reserve responded to the call. On foot, by motorcycle or by *chiva* (a colourful traditional Colombian bus), they came to the meeting point at the headquarters of the local Indigenous Authority, the Indigenous *Cabildo* of Canoas. An eagerness for action was palpable on that crisp morning. The participants wanted to work, to stop mining in the nearby Indigenous Reserve of Munchique. But why?

At the heart of Nasa discourses on resistance to mining are moral principles of Unity, Land or Territory protection, Culture and Autonomy. These principles resonate across the Cauca Indigenous organisations grouped under the banner of the Cauca Regional Indigenous Council, widely recognised in Colombia and internationally by its Spanish acronym, CRIC. The Nasa have been exposed mostly to illegal mining of a predatory and violent kind, an industry that has grown uncontrolledly in Colombia, leaving social and environmental degradation in its path. They do not hold large-scale mining in high regard either. In general, Nasa perceptions of large-scale mining are negative. As a result, the prevalent view amongst the Nasa participants of this research, both leaders and community members, is that as mining enters the Nasa territory, it clashes with their moral principles, threatening the Nasa world. To explain this, the Nasa use a range of what I shall refer to as 'narratives of mining-induced degradation'. Unity is a moral principle, but livelihood dilemmas divide communities between supporters and detractors of mining. Instead of protecting life and land, mining pollutes and destroys them. With mining come cash, weapons and violent groups that influence young people, weaken agricultural and spiritual practices, and displace traditional authorities. These see culture and autonomy begin to fade away. The list of grievances is long. In sum, many Nasa associate mining with social, environmental, cultural, economic and organisational degradation.

At the time of the working bee against mining, the Canoas community had experienced these very issues. Just a couple of weeks before, one of their own died in a mining-related dispute between Indigenous people in the Munchique Indigenous Reserve. Some members of the Munchique and Canoas Reserves were reportedly mining in Munchique territory. All of them were ignoring the Indigenous Authorities' ban on mining. Mining effectively divided the communities. It saw the Indigenous Authorities overlooked, and brought violence and the unfortunate death of a young person. And this was not the first time. Accounts by several people from Canoas spoke of land degradation and loss of livelihoods, unexplained deaths, family violence, violent groups threatening leaders and community groups, community division, miners' neglected children becoming easy recruits for guerrillas and other violent groups, and other social and environmental ailments.

Following the violent death of the young miner from Canoas, the Indigenous Authorities of the North Cauca Zone held an assembly to discuss the tragedy. Assemblies like this, about issues that concern communities, result in decisions that the community expects to see implemented. At that assembly, they decided to come together to stop mining in Munchique. The Authorities of Munchique and Canoas were not alone. CRIC backed them, and so did the Association of Indigenous Authorities of North Cauca (the Spanish acronym is ACIN), the predominantly Nasa sub-regional Indigenous organisation, affiliated to CRIC. The intercommunity division had rung loud alarm bells of a possible rupture in the Indigenous movement. The Indigenous Authorities convened the *minga*, or working bee, to perform a large-scale action, with the presence of external observers, including United Nations' Human Rights Commission delegates, journalists and researchers like myself.

Once at the meeting point in Canoas, the turnout was significant but not dramatic, perhaps 200 people, very motivated to act; that is, to go to Munchique to help in the minga. But there was confusion as to the next steps. Later the *Gobernador*, the indigenous reserve's highest authority, passed on the message that, for the minga to be effective, the technicians from the Regional Attorney's Technical Body recommended blocking the mining tunnels for good, using explosives. Otherwise, illegal miners could rapidly reopen the tunnels. Using explosives would require explosive transportation, licences and procedures, as well as an environmental study and management plan, all to be either coordinated or approved through mainstream State institutions. So, this Indigenous movement that takes pride in being nimble in acting on key causes was struck by a small episode of paralysis, as it faced a task far more technical than could be achieved with the shovels and machetes they had been asked to bring on the day.

The audience at the Canoas Indigenous Authority, or *Cabildo*, headquarters did not receive news of the technical delays with enthusiasm. Many had put aside the day's activities to participate in the working bee. Some decided to go back to their plots of land or other places of work, others encouraged the *Gobernador* to lead the community into Munchique to participate, regardless of the technical issues. Overall, there was a feeling of dissatisfaction and some voiced concerns about autonomy, one of those four fundamental moral principles. People were questioning whether the Indigenous Authorities were exercising their territorial autonomy. Or did they now need 'approval' from others to control mining in their own territory? The latter was not a welcome alternative for some. After all, the Indigenous movement in Colombia went through a long struggle to achieve its autonomies. The discussion at Canoas went on, seeming to stagnate, and as time started to get away, my Nasa host and I decided to join the working bee in Munchique.

The motorcycle ride to Munchique took over one hour, up through the green mountains of Canoas, until we reached just over 2,700 metres altitude in the Nasa settlement of La Aurora (The Dawn), in Munchique. Along the road, my host showed me the way to the gravel mine, owned by the Canoas Indigenous Reserve, or Resguardo, and used for construction materials. The mine was closed at the time, but was considered legitimate, because it was community run and used for local road maintenance. I was excited to return to Munchique, I was due to complete a stay there in 2016, with collaboration of the Munchique Cabildo Authorities, but community conflict about mining led the Authorities to recommend that I delay my visit. Now was a much better time to come. With a large group of people from different places in the region to talk and act on mining, I would not be stirring uncomfortable discussions about a difficult subject, and I could just join the many external observers.

We arrived in Munchique and, as usual in these Nasa gatherings, the anthem of the Indigenous Guardia – a widely recognised form of Indigenous civic guard that I shall talk about in Chapter 4 – was playing. It was an inspirational chant that placed the day's action against mining within the broader context of Indigenous resistance processes in Colombia and Cauca. The anthem told us about the strength and courage of the Guardia, about their struggle to survive and to attain justice, and about their peacebuilding efforts referring to the Guardia as "*friends of peace*" – the Guardia have long been acknowledged as a peacebuilding group.[2]

Rather than the thousands expected, there were 500 people or so at the meeting point. Why was the turnout lower than expected? I wondered. Some ventured that word had spread about the delays with the explosives, licences and technicians, and that this had discouraged more people from travelling to

Munchique. Others said that there was division in Munchique and the community might not act decisively to close the mines. Others mentioned threats from the Munchique miners, that they would leave traps in the terrain, endangering minga participants to fall into an open tunnel. In any case, the Cabildo of Munchique gave the go ahead, and people from CRIC and its North Cauca division, ACIN, spoke in their support. The latter two explained the technical delays. The Indigenous Authorities agreed that those present would go up the Cerro Munchique Mountain to block the mining tunnels, despite knowing it to be only a temporary measure. The people had come, some after long trips, and they could not just be sent back.

There were all sorts of people at La Aurora. Indigenous Guardia, United Nations (UN) representatives, mothers with their small children, journalists both Indigenous and foreign, Traditional Authorities, Traditional Doctors, Indigenous people of all ages, and a big contingent of cooks who would prepare lunch for the tired minga participants. We lined up to see the Traditional Doctor, a community Elder, for a harmonisation ritual, each one of us in turn, ahead of the walk. He poured water with plants over people's heads. His instructions came in NasaYuwe, the language of the Nasa. My Indigenous host did not speak it but the Traditional Doctor repeated the same words impatiently. Finally, frustrated, the Elder said "*This one! He still doesn't speak NasaYuwe!*" When my turn came, he said some words in NasaYuwe, and told me to stick to my host, all the time, not to venture away from him.

Mining tunnels dotted our path up the mountain of Cerro Munchique. I counted about seven along the brisk walk, but concerned with keeping up, I must have missed some tunnel openings. I could not help but wonder who would rehabilitate all this land, where the funds would come from, and how people would handle the safety risks from those holes in the ground. Then a passer-by brought me back to the moment complaining that the Guardia did not seem enthusiastic about the working bee. It was hard for me to discern what this lack of enthusiasm might entail. But shortly after, I was thoroughly convinced that energy was not in short supply. I lost sight of my host and of everyone else for a brief moment. So, I raced through a thickly forested area, following the voices ahead. I found my host and the others, across a big trench that I was reluctant to cross, but it seemed there was no other option. No sooner had I taken the first step forward than the others shouted "*No, No, No!*" So, I stopped, only to hear a loud creaking sound, followed by a crash as a tall tree landed a few metres in front of me. I was glad that I had done the ritual with the Traditional Doctor; I also regretted having looked at my phone which was what let me lose track of my host. In any case, chopping down a big tree to block a mining tunnel did not strike me as a lack of enthusiasm. It seemed paradoxical. I felt sure there was a background story.

The light blue vests of the UN observers stood out against the red and green scarfs and flags of the mostly Indigenous crowd. Their strong presence was a stark contrast to the absence of some State institutions. Actions like the working bee seek to exercise and protect Indigenous law and autonomies. The presence of UN human rights delegates was seen as a step towards guaranteeing the rights of the Indigenous people. With the UN also came a Canadian journalist, who was eager to clarify that he was not one of the 'bad' Canadians, as he quickly volunteered that it was Canadian consultants who wrote the Colombian mining code that had, according to him, and to several others I encountered, caused so much trouble to communities, by putting corporate interests first. The UN representatives and the journalist were important actors in an organised effort to control and raise awareness about a type of mining that had turned violent. They exemplified the collaboration and communication work the Nasa embed in their resistance to mining. But one element of the collaboration effort was glaringly absent: the explosive experts from the Regional Attorney's Office. Either the State[3] bureaucracy was unable or unwilling to keep up with the fast pace of Indigenous mobilisation, or the Indigenous Authorities were not adept or realistic in navigating State Institutions and regulatory requirements, or a combination of all. Nevertheless, the result was not ideal for the minga and its participants. The question, as far as this collaboration was concerned, revolved around the interplay between Indigenous territorial autonomy and Authorities, and mainstream authorities and regulatory processes run by the State. The Nasa navigate this tension whenever they deal with mining, because there is a struggle between mainstream Government and Indigenous Authorities over territorial management, in particular the management of subsoil resources.

Despite all the activity, the discourses, the music and the people, there was a sensation of the doldrums hanging in the air, as if something remained untold, and minga participants seemed to be deliberately avoiding any conversation about mining. The prevailing moral and institutional discourse was one of Indigenous resistance to a kind of mining that was driven by external, powerful and not always benign forces. However, in practice, in Indigenous territories, there were also Indigenous miners. At times, these Indigenous miners were the unlikely frontline of an illegal industry, managed by corrupt actors, that multiplied throughout the Colombian territories triggering a wide range of conflicts and deleterious effects. At others, it was the Indigenous Authorities themselves who had decided to take mining into their own hands, inspired by the ideal of a socially driven mining keeping foreign companies at bay, or simply because communities needed basic construction materials. Still at other times, Indigenous mining entrepreneurs were born of individual initiative amid scarce economic opportunity. They sat awkwardly in a context of moral opposition to mining. Indigenous mining in Indigenous territories was contentious, hence the air of stagnation on the day of the minga. There were a range of views on mining and because of that there was nuance, uncertainty and caution when dealing with it in the public arena.

So, we walked, first up and then down the sacred mountain of Cerro Munchique in what I would describe as a powerful symbolic endeavour. As the air grew thinner on the ascent, the mining tunnels became more conspicuous, and some of us grew more uncomfortable: the sacred mountain of the Indigenous people was full of mining tunnels. As we walked down, there were comments coming and going: Was this a futile exercise? The tunnels could be opened again, any time, and mining would continue. Regardless, the minga was no doubt an event of great symbolic and political importance. Only the explosives experts, that is, outsiders, could close the tunnels permanently. However, a public act of opposition and of moral condemnation had been performed, placing the miners and the violent party as trespassers of moral and legal mandates in front of the entire Cauca Indigenous movement, State authorities and the international community.

The day at the minga encapsulated many of the contradictions, tensions and urgent questions that participants from the Nasa community and elsewhere had discussed with me during my time in Cauca, Cali and Bogotá. In the confluence of strong economic processes, moral principles, aspirations and power relations, the tensions were palpable. For some, the region was set on a path to growing violence and illegality. Others, however, still had hope. Why was there a sense of pessimism in some, and a determination to exercise resistance in others? For the former, perhaps the determinant was the experience and memory of colonialism and dispossession, of armed conflict and illegal economy-driven violence, of disorganised and unfair administration of wealth, including mineral resources. For the latter, the resistance and the fight for Indigenous rights and autonomies were imperative. It is necessary to refer to the backdrop of these complicated tensions to set out the context where the Nasa community's responses to mining take place. However, it is in the workings of Nasa communities and organisations that we can observe why, to many Nasa, responding to mining is effectively building peace. It is in this space that my inquiry concentrates.

Argument: Nasa Responses to Mining and Why They Matter

Why explore Nasa responses to mining during the post-conflict transition period? Nasa influence on community discourses on mining shall echo far beyond the Cauca region and Colombia, in particular among Indigenous peoples. The way the Nasa configure a response to mining will likely influence, inform and reflect other responses to this economic activity in the Latin American region. The Nasa are a reference point on organised community resistance not only in Colombia but in Latin America (Gonzáles Rosas, 2016), a region where Indigenous actors have gained prominence in national debates (Bebbington et al., 2018). Nasa communities are experiencing the fears and tensions characteristic of post-agreement transitions in Colombia. The mining-related questions and difficulties that both ACIN and Canoas faced during the uncertain times of peace agreement negotiation are relevant to Indigenous people and other communities in conflict, post-conflict and peace scenarios.

I shall examine the Nasa Indigenous people's responses to mining in the Colombian post-agreement transition between 2015 and 2018. I explore the customary moral frameworks that inform community views and responses to mining, ideas of peace, and peacebuilding activities. In collaboration with Nasa organisations from North Cauca, I conducted a layered ethnographic case study, exploratory and interpretive in nature, building on archival research, interviews and participant observation at various domains, including a Nasa sub-national association, namely the Association of Indigenous Authorities of North Cauca (ACIN), and the Canoas Indigenous Reserve, a Nasa Indigenous territory affiliated to ACIN.

I place community responses to mining in post-conflict Colombia in a position of centrality, recognising the agency and sophisticated organisational processes of the Nasa of North Cauca. Often studies of post-conflict natural resource governance have sought to prescribe community roles and steer community actions in the direction that the international peacebuilding community considers appropriate, construing communities mostly as passive (Chapagain & Sanio, 2012; Reimann, 2004). The story of the minga in Munchique is but one example of how the Nasa are far from passive. They are part of an Indigenous movement of colossal strength, with a number of autonomies recognised in the Colombian Constitution (1991), active in the peacebuilding arena, with a sophisticated tradition of social mobilisation, whose public discourse expresses direct moral opposition to the growth of large-scale or other forms of high impact mining operations in Colombia. It is necessary to ask: What can we learn from what communities are doing, from their own questions, tensions and difficulties in addressing mining-related problems in a post-agreement transition?[4]

I contend that while the prevalent Nasa view was that mining causes moral breaches and risks Nasa lives and aspirations, there is more nuance and tension when it comes to miners that are members of the Nasa community. The Nasa moral economic thinking delineates an organic perspective of the economy where correct actions equate to life-sustaining reciprocity and transgressions sever vital reciprocity. The Nasa ideals about a good life and their moral principles contrast with the mining-induced moral and economic shifts in their territories and underpin the NO to mining mandates. There are internal tensions because material need and an urge to protect Nasa territories from external miners have seen the emergence of Nasa miners, independently or with the support of their Nasa Authorities. Those who do not oppose mining ask for a pragmatic test of the NO position in light of material circumstances, and economic, legal and environmental dynamics. Being a Nasa miner or mining territory is marred with moral contradictions that persist despite efforts to elucidate a coherent moral discourse.

I found that many Nasa observed strong links between an expanding mining frontier and violence and repression against them, which mobilises them to resist these forms of mining despite limited government support. They resist through practices consistent with the Nasa tradition of resistance to the Colombian armed conflict, and with the resistance practices of other Latin American Indigenous peoples. Nasa responses to mining are part of an institutionalised, everyday resistance. They include a legal dimension, an institutional design, specific projects, territorial control and enforcement, as well as research, dialogue and collaborative advocacy that are far from uniform, and instead involve local nuance and interpretation. This multifaceted and multilayered resistance consumes significant community resources at the sub-regional and local levels. In contrast, the Colombian Government had failed to halt illegal mining in Nasa territories or to support Nasa community mining ventures.

Exploring the links between the Nasa responses to mining and peacebuilding work, I observed that where mining by outsiders is concerned, there is broad consensus that resisting mining equates to building peace. There is a prevalent view among the Nasa that there are connections between mining and illicit economy networks with links to violent armed groups. However, livelihood dilemmas, aspirations for better material conditions and political contests see some Nasa participate in unsanctioned mining ventures in Indigenous territories. In these scenarios the Nasa are divided on what constitutes peacebuilding where responses to mining are concerned. Ultimately, violence creeps in through the cracks left by dispossession and material poverty while the Colombian society fails to attend to those problems in rural territories such as North Cauca, so that communities' best efforts fall short. Nasa organisations find themselves filling out the gaps that other peacebuilding actors leave. The Nasa need national and international actors to respond to the interactions between mining and armed conflict through support for the Nasa's peacebuilding work and with complementary actions coherent with the Nasa's lived reality.

Having outlined the argument, I shall now describe the collaborative research approach this work is based on, the idea of morality as the lens I applied and the structure of the book.

Collaborative Research Approach

In this book, I summarise four years of work, including the learning that emerged from nine months of fieldwork spent between North Cauca, Cali and Bogotá, foregrounded by previous work with the Nasa on mining in North Cauca. With the collaboration of ACIN and the Canoas Indigenous Cabildo and Community, I conducted a layered ethnographic case study of the Nasa response to mining and its links to peacebuilding. The fieldwork began at ACIN where I conducted ethnographic observation basing myself at the headquarters of the organisation and participating in activities connected to peacebuilding and responses to mining. I also conducted interviews gathering narratives and information from 17 people from all levels of ACIN, 27 per cent of whom were women. In collaboration with ACIN and Nasa Indigenous reserve authorities, we identified the Canoas Indigenous reserve as a case study. I interviewed 17 people in Canoas, including 35 per cent female interviewees. In addition, I conducted six interviews in other Nasa Resguardos that had observed mining dynamics in their territories, and over 20 interviews with people from outside the Indigenous movement. After analysis and debate, I presented the preliminary and final findings to ACIN and the Canoas Authorities for feedback. The results of this process constitute this book. I refer to Nasa interview participants by pseudonyms of their choice, and to interviewees from outside the Nasa communities by their institutional affiliation, where authorised.

The idea for this research emerged in the context of a previous research collaboration with ACIN about mining governance. ACIN proposed we studied the intersection between mining and post-conflict dynamics in Nasa Indigenous territories. This request led me to focus my PhD dissertation on Nasa responses to mining in the post-agreement transition. The Colombian Government and FARC had been negotiating an agreement for years; the end of that armed conflict somehow seemed within sight to many of us, and the post-conflict scenario was the next logical frontier to examine. There was huge uncertainty about what would come next and fears about what would happen to Indigenous territorial control and autonomies in the post-agreement scenario.

I shall share my recollection of how we developed the research design and implemented the research. The responsibility for the academic output was in my hands, I undertook all interviews, data collection and detailed data processing and analysis, and I am responsible for the analysis presented in this book. However, this was a collaborative exercise and an ongoing dialogue with Nasa organisations and researchers. Mining is a contentious economic activity in Nasa communities. So, from the perspective of interviewees and Nasa authorities, it was best for an outsider like me to conduct the interviews without local community members present to allow people to speak freely in interviews and avoid tensions. Nasa researchers and activists who became my collaborators gave input into the design, identified activities I could participate in and people I could interview, gave feedback on preliminary analyses, vouched for me to support my access, and defended the research idea within their communities and organisations.

Research Agenda

After the request from ACIN to conduct research in the intersection between mining and the post-conflict in Nasa territories, I iteratively drafted proposals for ACIN to consider and revised them following feedback. Via email from Australia, I would communicate with a member of ACIN's Planning Team, Marco Fidel Mosquera, and with a prominent Nasa political leader, although over time the communication became solely with Marco. Bringing together the input, what I learnt in our previous joint work, ideas from readings and in discussions with my thesis advisors, a specific proposal for a research project took shape in a detailed description document. It included general information for participants such as the research aims and activities, and how I would use the information people shared. It dealt in detail with matters of import to the Nasa such as protecting participants' identities, the processes to present drafts to Nasa organisations and for them, or individual participants, to give feedback, and what specific output the Nasa organisations would receive. Marco would talk to colleagues about the proposal and share suggestions with me. Once ready, Marco discussed the proposal with the Tutenas of ACIN, their highest authorities. We agreed I would meet a group of ACIN representatives to present the proposal face-to-face to them for their consideration. Being based in Australia, I travelled to Colombia in the hope that the meeting would result in an agreement to proceed with the research. In preparation, I produced a narrated animated presentation that participants could see in advance of the meeting, or in its place, should they not be able to attend.

We held a meeting in mid-July 2016 at the ACIN headquarters in Santander de Quilichao, North Cauca's main town. There were at least five ACIN representatives, most of them leaders of programmatic areas at ACIN, known as *tejidos* or networks. Another two ACIN leaders were absent but available to meet or talk later. We gathered around the large meeting table at the office of the Planning Program, did our introductions and I presented the proposal. Following this, each of the ACIN representatives, all of them men, offered comments and feedback. I was expecting an affirmative response, given I had been asked to have a face-to-face meeting and they knew I lived very far away from Colombia. What I did not expect was that the participants would be interested in a study much larger than what had been suggested in earlier conversations. They were interested in me doing detailed case studies of six different localities in addition to ACIN. The ACIN representatives explained to me why those six Cabildos were important, the type of mining taking place there, and the complexities associated with mining in those territories. The cases were diverse, ranging from formalised small-scale Indigenous mining, to outsider illegal miners that local communities and authorities opposed, to abandoned Indigenous mines, to Indigenous people mining gold despite a local ban on mining. The array of cases was a fascinating prospect. At the same time, seeking Free, Prior and Informed Consent (FPIC) from six different Cabildos and from ACIN, and navigating relationships with actors connected to mining in such a diverse range of situations implied a larger project than a PhD could cover. The risks to navigate would become more complex with more actors and territories to cover. So, I offered to focus on ACIN plus one or up to two Cabildo cases, and to seek interview participants from all the six localities, subject to FPIC from the Cabildo Authorities and ACIN Tutenas. The ACIN team agreed. I kept a detailed record of the meeting and circulated minutes requesting any corrections, but none were made. Based on our previous research collaboration, I had proposed to have a liaison committee to provide advice on the organisational and territorial processes and help establish relationships. The ACIN team had decided that Marco and one of the people who were absent were to be the liaison committee members.

Shortly after, I met Luis Omar Collazos, an economic-environmental specialist, who would be the remaining liaison point at ACIN.

The next step was to document the consent from ACIN. Marco had discussions with the Tutenas to foreground my meeting with ACIN's legal representative who was also a Tutena. Earlier, I shared a draft proposed letter of consent for ACIN to consider and adapt. Marco shared it with the Tutena while I was in Australia. I met the Tutena at an agricultural product distribution cooperative owned by ACIN. She was well informed about the project and Marco gave her some details on the spot to remind her. I explained the project briefly and described the months long discussions and feedback as well as the face-to-face meeting held recently with Marco and others. I stated that I was seeking FPIC to proceed with the permission of ACIN. Marco produced a folder with the consent letter in ACIN letter head and gave it to the Tutena who gave it a final read and signed it in front of me. She asked Marco to make sure to keep a copy on file and do what he could to help me. A precious document to formalise our agreement. The letter explained what I would offer to ACIN, what ACIN would offer to me, and what permissions they had granted me.

I was fortunate that some ACIN and Canoas leaders and researchers interested in seeing collaborative research were generous in offering guidance, support friendship. Marco Fidel Mosquera, Nelson Valencia and Rosalba Velasco were vital to the research. Marco guided me through the process of negotiating access to ACIN, identifying people and documentary sources and gaining access to various territories. Rosalba Velasco and Nelson Valencia offered their insight into the Nasa collaborative research approach, the local politics and the dynamics of mining, and together with their son, Daniel, hosted me in their Canoas home like they have done with several other academic researchers. Rosalba had led ACIN's former research unit or think tank, *Casa de Pensamiento*, and taught me about their collaborative research where external researchers had an ACIN-financed co-investigator and the research agenda was negotiated to ensure local relevance. With Casa de Pensamiento gone, there were no funds for me to have the privilege of a salaried co-investigator. However, I was aware that this would be a collaborative effort. Nelson and others had already researched mining and environmental matters in Canoas. Much to my fortune, Nelson took it upon himself to help me liaise with the Canoas Cabildo and gathered a local group of activists and researchers interested in the project.

Resguardo Case Study Selection, Design and Set-Up Process

I sought to visit the six Nasa Indigenous Reserves that ACIN personnel had suggested as case studies. The Resguardos were Delicias, Corinto, Canoas, Munchique, Tacueyó and Huellas. Marco would either give me contact details or put me in touch with the Cabildo or with local leaders or researchers who could help me contact the Cabildo. Contacting people was a prolonged process because the Nasa leaders receive threats from violent groups and are not prone to answer the calls of unknown numbers, and their rural journeys or residences did not always allow access to mobile communications networks. Gradually, I was able to speak on the phone with the Cabildo Governadores, or their deputies, or their environmental-economic specialists. In Canoas, I was in touch with Nelson Valencia and Rosalba Velasco from the start, both seasoned researchers whom I had met through previous research work with ACIN. Nelson helped me enquire about the project in Canoas. Over time, I received permission to visit each of the Resguardos, in several cases also with an offer of Indigenous Guardia protection to access the territory. The Guardia of Delicias, Munchique and Canoas met me at main road intersections and escorted my vehicle into the headquarters of their Resguardos to make it evident that I was visiting the Cabildo. In the case of Tacueyó, ACIN sent one of their employees to accompany me in the journey travelling in my car, should there be any questions asked along the road.

At each of the Cabildo headquarters, we held meetings. Some were with a couple of people, some were with the full Cabildo of up to 20 or 30 Indigenous Authorities, or with a large assembly comprised of the Cabildo and members of the Guardia and the broader community. Sometimes, the meeting was exclusively to talk about the potential research. Other times, the Cabildo allotted me time to speak at their regular meeting, among other agenda items. Participants would ask questions and make suggestions about the research. I would take notes, offer clarification and explain what I could and could not offer. At Canoas, Delicias and Munchique, Traditional Doctors met me and gave their opinion as to whether I "gave them good signals" *les daba buena seña* – essentially their observations, based on their spiritual and moral authority and knowledge, on whether I was to be trusted or not. After the process, it became clear that the Cabildos with most interest in participating were Canoas, Delicias and Munchique. I continued to follow up with their Authorities, made at least two more visits to each for further discussions and conducted six scoping interviews in those territories.

Based on Cabildo consent and celerity, and on the support of Nasa local experts, we set out to do a case study of the Indigenous Reserve of Canoas. Canoas is in the rural area of Santander de Quilichao. Some areas of the Resguardo are only 20 minutes from Santander de Quilichao's urban centre. Canoas has a very well organised community with a strong political and environmental conservation tradition. Some of ACIN's most prominent leaders come from or reside in Canoas – Feliciano Valencia, a former Senator, and Rosalba Velasco, who held the post of legal representative of ACIN, are some examples. Canoas has a seasoned group of environmental researchers and activists. Among them not only Rosalba Velasco and Nelson Valencia, but also traditional law expert Rodolfo Pilcue; and emerging leader Katherine Muelas-Martines. Their interest in the research helped facilitate discussions with the Canoas Cabildo and with the broader community. They saw value in the research and vouched for me at community assemblies. This is how after three visits to the Canoas Resguardo, the Cabildo gave their consent, and Canoas became the Resguardo case study.

Nelson Valencia convened a group comprising also Rodolfo Pilcue and Katherine Martinez-Muelas that became the liaison team for the case study. We held a planning session at the Vilachí Hacienda or Harmonisation Centre.[5] At that meeting, I gave a detailed account of the research as it had been designed in collaboration with ACIN and asked for the group's input. They offered plenty of ideas about who could be interviewed, what I might find out in interviews and the key localities to cover. The group gave me guidance on the process to report back on findings and seek feedback, and on the best ways to approach potential participants. The team thought that an outsider would be best placed to interview people about mining because it was a polarising topic within the Resguardo. Likewise, they felt that a report from someone from outside the Resguardo would be seen as more credible, in particular if the people knew that it was done with the support of the Cabildo and the guidance of local researchers.

Implementation

At ACIN, Marco proceeded with introductions and helped me identify documentation for review and interviewees. Marco and Luis were my guides regarding interviewees. Marco introduced me to the Tutenas as we found them during my days at ACIN or other activities in rural areas and recommended that I interview some of them. The early mornings were a good time to meet people, they would come in a salutatory mood, before the day got too hectic. The early mornings were also when people were congregating for the day's journeys and sharing more philosophical reflections. It was in the early morning routines of getting ready to depart for rural activities that I got to meet several of the Tutenas and interview participants.

During my time at ACIN, I became familiar with the rhythm of everyday organisational life and witnessed some key events. I sat at the office of the Planning Program in Santander de Quilichao, went to ACIN's rituals and peacebuilding and mining-related activities, interviewed their personnel, and consulted their documentary sources. The office of the Planning Team was also the main meeting room at ACIN, and the planning team would remain there as other ACIN staff and leaders held meetings. I was allowed to stay there during meetings. When not dealing with interviews, I would read or write fieldwork notes while at the office, so these meetings became a powerful source of context information, names and anecdotes that I could learn from. In that room, I observed the reaction and mood after some of the setbacks of the peace process, such as the results of the 2016 plebiscite where the NO won by a narrow margin. I witnessed internal discussions on Indigenous strategies for the post-agreement transition, as well as numerous internal training and debate events about the Peace Agreement. Outside the ACIN headquarters, I observed larger events and mobilisations that ACIN led, or participated in.

For example, I saw ACIN taking part in a large-scale march for peace in Bogotá, accompanied them in their "walk through the territory" of the Delicias Indigenous Reserve, where we walked from farm to farm, door knocking and Marco talked about ACIN and the importance of the Peace Agreement, and I spoke of the research I was doing on mining. I also witnessed an Office of the High Commissioner for Peace workshop about the Peace Agreement for ACIN leaders and practitioners held in the rural area of Santander de Quilichao at the Munchique Resguardo's El Gualanday property, and another similar event in Puerto Tejada's urban area.

I went to Canoas to spend two weeks doing interviews. The liaison team put together a list of potential interview participants. Nelson Valencia helped me arrange the first interviews and I arranged the rest. I arrived with a member of the liaison team to all interviews. The commute was a wonderful opportunity to listen to narratives about the landscapes we were traversing. Following the advice of the liaison group, and given there was confrontation within the Resguardo about mining, I did the interviews on my own, and reported back findings in de-identified format to allow people to speak freely to an outsider. Throughout the analysis, I was in regular communication with the local researchers and activists I was collaborating with, in particular Nelson Valencia in whose home I was staying. This communication was fundamental in ensuring that the research remained connected to community questions and concerns.

Both Canoas and ACIN received preliminary and final findings reports in Spanish that I delivered verbally in person to their leaders in sessions where I also requested feedback. All members of liaison groups received draft written reports in Spanish for comment. The participants who provided an email received a final findings report. Nelson and Rodolfo took the hard copy of the report to the Canoas Cabildo for their formal records. Marco did the same at ACIN, where Rosalba, by then their legal representative, also received an official copy.

The authorities of both Munchique and Delicias had expressed strong interest during my visits to their territories and I was extremely close to embarking for Munchique. However, the conflict about mining in the Munchique Resguardo led the Cabildo to ask that I postpone my visit just two days before I was due to depart. It was not the right time for the research because the situation was tense. In Delicias, the authorities were keen to have a case study done about their Resguardo, and I visited Delicias three times to discuss the prospect and take part in other activities. Unfortunately, I ran out of time and funding before they could give a definitive response. During my last week, one of the Delicias Authorities asked me if I could come to do the case study, but by then I was due to return to Australia. I felt sad to miss this precious opportunity to work with them and hope one day to return.

Research in a Scenario of Conflict and Violence

Knowledge on Nasa responses to mining can become unavailable because of the complexities that accessing field sites and sustaining research in Nasa territories represent. More than five decades of armed conflict, and the intense levels of violence resulting from the conflict and the illegal drug and mining economies that finance it, make access to North Cauca Indigenous territories difficult and complicate conducting research. Cesar Rodriguez-Garavito (2011) coined the term 'minefields' to refer to these sites where the capitalist resource export-oriented economies converge in Colombian Indigenous territories. "*They are minefields*", he argued, "*because they are highly risky; within this terrain, social relations are fraught with violence, suspicion dominates, and any false step can bring lethal consequences*" (Rodriguez-Garavito, 2011).

It was possible for me to apply the methods I outlined in the previous pages because I am a Colombian national and a native Spanish speaker,[6] from a different region of Colombia but with deep familiarity with the Cauca region, where I first worked in 1997 and 1998, and later with Nasa organisations through research collaboration from 2013 to 2015. I had the linguistic competence and context familiarity to travel in Cauca. Many Nasa thought of me as both a collaborator, which allowed access, but also an outsider and a 'white woman' – perhaps because of my long-term residence in Australia – who needed explanations. I grew up during the 'war on drugs' and some of the most intense moments of Colombian armed conflict. Like most Colombians, I have victims of the conflict within my family network and my family was once an internal refugee *campesino* (peasant) family. Because of and despite this, I do not see the armed conflict and the violence as protagonists but as important elements of context that one must navigate. I shall speak about the violence and the conflict in Chapter 2, to contextualise, but shall otherwise not centre the narrative on conveying a sense of the danger and the violence that surrounded the communities I worked with. This has been done extensively in the seminal work of Michael Taussig (2003), for example. To attempt that here would only distract from appreciating the ideas and the work of the Nasa that I wish to highlight.

In a similar vein, I do not consider the armed conflict per se a limiting factor to the conduct of this research. This would be contradictory, because the conflict is a central contextual factor to the research. When I speak of navigating competently the context of the armed conflict, I refer to considerations that inform all aspects from research instrument design to logistical aspects. I would like to illustrate this with an example – the armed conflict context required that I enter rural Indigenous territories accompanied by members of the Nasa community. This resulted in a significant negotiation and logistical effort. This is no different to working in other Indigenous communities, based on my experience in Australia and Papua New Guinea. Nevertheless, it allowed me better access to local leaders, individuals and organisations, because evidence of a due process of consent and collaboration foregrounded my arrival. It meant I had to exercise prudence with the length of my visits to Indigenous Reserves, so as not to become a burden, and as a result I received focused attention locally and people made an effort to be available to me, being aware of the efforts behind the visits. They were willing to speak with me because I was accompanied by members of the community. This was regardless of the disagreements that might have existed between my official hosts and some of the research participants, which often came up in the privacy of interviews. In many cases, I was asked to interview people from different sides of the mining debate, who had been publicly silent, so that this research could, in the future, inform better community understandings. By navigating what could be seen as 'restrictions' in this armed conflict context, I gained further insight about the dynamics of the conflict and about trust building in Nasa organisations and territories in North Cauca. Last but not least, I had extended access to local people interested in the research – my hosts – and it should not come as a surprise that they held substantial knowledge about mining dynamics in their territories. Because of the Nasa's and my own competence in operating in an armed conflict environment it was possible to access the research site and collect data. In absence of that competence, the armed conflict would have indeed become a limiting factor and a barrier to research.

Morality

Morality is a central fault line in the conflicts I address in this book, as it usually is in class-based conflicts and conflicts surrounding high impact economic activities like mining. Discussing what I observed in the Nasa communities in this study requires addressing questions of morality. In doing so, I draw on the concepts of moral economy and Living Well (*Buen Vivir*). These allow direct engagement with the most contested and therefore conflict-relevant aspects of the mining debate in the Colombian post-conflict transition.

Let us first attend to the notion of the moral economy. The moral economy is based on understanding customary ideas and consensus about how market activity should work (Owen, 2009). Thompson's (1971) seminal study of food riots in England in the mid-eighteenth century addressed the shared understandings and principles that those living in materially poor circumstances held about how grain markets should work at that time in history. Thompson questioned the idea that the rioting crowds were an irrational hungry mob. Instead, he contended that they had shared moral principles about the grain markets and rioted when food prices were too high, because this was an indication that the moral economic order was broken (Thompson, 1971). Thompson is the most prominent of several historians including Hobsbawm (1971), Rudé (1964), Hilton (1973) and Hill (1975) who studied European resistances throughout time, from the perspective of subaltern classes rather than the rulers. They illustrated the ritual elements (Hobsbawm, 1971), concrete achievements (Hilton, 1973), diverse moral and pragmatic motivations (Rudé, 1964) and prevalent ideas and beliefs (Hill, 1975) that were part of those resistances, including a strong levelling out tendency and desire for social justice.

The concept of the moral economy has been used in a range of studies and is pervasive in the study of social mobilisation, resistance and riots in the global south, including protests and rebellions in Latin America (Siméant, 2015). However, the moral economy reaches far beyond riots and other social mobilisations. They are one element of the moral economy, but not the whole. Riots and social mobilisations are manifestations of discontent that seek to adjust the market to respond to that discontent. Scott (1977) referred to the moral economy as embedding ideas of economic justice and exploitation, and used it to examine peasant resistance to capitalism in South East Asia. Smaller scale, every-day forms of resistance, such as those described by Scott (1985), also constitute manifestations of the moral economy. It is ultimately in the principles, concepts and customary understandings that people hold about the economy, in the moral economy, that we find the origins of those forms of resistance large or small.

There are different interpretations of the concept of the moral economy: an institution, a set of intellectual tools or a disciplinary area of its own (Owen, 2009). It can be interpreted as the architecture of shared principles, mechanisms, rules and norms that regulate economic relationships (Sandberg, 2015; Siméant, 2015); or as the way this architecture becomes institutionalised in social practices and a way of life (Sandberg 2015); or as the social embeddedness of economic activity (Siméant, 2015). Here I will follow Owen (2009) by understanding the moral economy as an intellectual device. This will place morality as an important element of this inquiry.

Many thinkers have explored different moral understandings of markets and people's role in markets and in nature. Examples include the moral economy of scholastic thinkers, Marx's analysis of capitalism, Polanyi's perspectives on the industrialisation of Europe (2001), Pope Francis's Laudato Si (2015), or Aristotle's notion of the good life, to name a few. Several of these thinkers or traditions have articulated the limits and purpose of our engagement with markets and with nature. Scholastic economics emphasised reciprocity and solidarity with its focus on justice, charity, the common good and prioritising human beings and nature over profit (Armstrong, 2016). Economic thinkers such as Marx and Polanyi were critical, as were the scholastics, of the tendency to turn nature sterile, while endowing money with the capacity to be fruitful (Armstrong, 2016). The first element in the critique speaks of the limits that our bond with nature should set, while critiques of capital as self-reproducing question accumulation are ultimately exhortations to attend to economic justice. Polanyi (2001) called into question how the natural environment, human activity and the medium of exchange could be turned into the "fictitious commodities" (p. 71) of land, labour and money, risking "annihilating" (p. xxv) humans and nature, while masking what in fact are intricate links between the economy and social relations. He illustrated the myth of unregulated free markets, and how unlimited growth and trickle-down economics not only do not work but also harm the poor (Polanyi, 2001). Marx likewise took issue with the accumulation resulting from money accruing interest, following Aristotle's similar analysis (Marx, 1992; Aristotle in Marx, 2000, p. 237). In Laudato Si, Francis connects environmental damage to a humanity that sees itself as "lord and master" overlooking its indivisible relationship with the nature of which it is made. Francis describes an "intimate" connection between the poor and the fragility of the planet when stating that questions of justice are part of environmental debates and that one must hear "both the cry of the earth and the cry of the poor" (2015). Aristotle's perspective has a more anthropocentric approach where it comes to understanding nature, which it sees as existing for the sake of humans (for example as described in *Politics*, 1948). Aristotle's notion of the good life should not be confused with the *Buen Vivir* ideal of Latin American Indigenous societies including the Nasa, which emphasises collective wellbeing and processes more than individual realisation or qualities.

The notion of Living Well or Buen Vivir, central in Latin American Indigenous thought, articulates moral ideas about the engagement of people in nature. It has been central to debates about extractive activity, development and Indigenous and other land-connected peoples in Colombia and more broadly Latin America. Buen Vivir refers to the reciprocal relationship between people and nature that promotes harmony amongst peoples and with nature (Acosta, 2008; Consejo de Mujeres Originarias por el Buen Vivir, 2015; Gudynas, 2011; Walsh, 2010). It takes issue with the idea of development and linear progress, focusing instead on their negative effects (Acosta, 2008; Gudynas, 2011). What Buen Vivir emphasises is knowledge, social and cultural recognition, ethical and spiritual codes of conduct about human–nature interaction, and aspirations for the future (Acosta, 2008). Proponents of Buen Vivir advocate it as a way out of a fraught extractivist model – where development is contingent on mass exploitation of natural resources – to post-extractivism, which connects ideas of post-development (Escobar, 1995), post-growth and de-growth (Acosta, 2010, 2016, 2017). Buen Vivir is not an isolated discourse and has engaged and included other global discourses on alternatives to both development and sustainable development (Gudynas, 2011). The Buen Vivir concept is not unified either. Instead, its various local interpretations have gained prominence in national debates in Latin America. Ecuador and Bolivia have enshrined Buen Vivir in their Constitutions. In Ecuador, for example, there has been ample debate as to whether Government practice abides by Buen Vivir ideas, amidst extractive expansion and the criminalisation of protest (Bebbington & Humphreys Bebbington, 2011; Walsh, 2010). However, a richer picture emerges from analyses of how various actors articulate and respond to the Buen Vivir discourse (Fitz-Henry, 2015).

A key motivation for North Cauca Nasa organisations' local peacebuilding and actions about mining is hope for a better life, free from the exclusion and violence of the prevailing political economy and armed conflict. Those Indigenous hopes for a better life reflect the regional Latin American Indigenous and tribal people's discourse of Buen Vivir. The concept brings a regionally relevant moral dimension to a political economic debate that has traditionally neglected morality. The armed conflict and mining conflicts embed contradictions between people's views about how the political economy, mining activity and markets should work. These views express a moral framework, or a moral economy.

Book Structure

After this introduction, I begin by outlining the multilayered context where this research took place. This merits a separate Chapter 2, which addresses the nature of the Indigenous organisations that participated in the research, the armed conflict, the dynamics of mining, and processes of colonisation, oppression and dispossession in Colombia. The subsequent Chapter 3 describes what I learned about the moral views of the Nasa organisations and individuals who participated in this research. This chapter shall deal with the prevalent view that mining clashes with Nasa moral principles. As this is not a uniform view, in Chapter 4, I highlight the perspectives of a number of Nasa participants who are miners or who offer alternative angles to the movement's official position. Chapter 5 is about the responses to mining I observed or was told about during my time with the Nasa, concentrating on the sub-regional domain of North Cauca. Later in Chapter 6, I analyse these responses to mining in the localised context of a Resguardo, Canoas, in Santander de Quilichao. With these as foregrounding, in Chapter 7, I describe the Nasa views on armed conflict-mining-peace interactions, including a discussion of the participants' notions of peace and peacebuilding and of the Nasa resistance identity that has shaped Nasa peacebuilding and responses to mining. I have chosen to place the perspectives from outside the Indigenous movement last rather than as a framing element for the Indigenous views. This is a deliberate decision to narrate in an order that places the Nasa at the centre, rather than as a peripheral position that contrasts with a notionally uniform national view or national interest in mining. Chapter 8 engages some of the perspectives from outside the Nasa community, for example from industry, academia, the Government, the peace negotiating institutions, international aid and cooperation institutions, and the campesino and Afro-descendant movements. This is followed by a concluding discussion in Chapter 9.

Notes

1 FARC was a Marxist-Leninist guerrilla organisation created in the 1950s in the Tolima department. Its origins are in Colombia's liberal party. The seminal event that triggered FARC's emergence was the National Government's bombardment of peasant organisation settlement in an attempt to crush community organisation and resistance. The armed conflict between FARC and the Colombian Government lasted over half a century.
2 The Guardia won the National Peace Award in 2004 (Redacción El Tiempo, 2004).
3 Throughout the text, I refer to State or to Government to reflect the following definitions. The Colombian State institutions have three core functions – judiciary, governmental and legislative. The legislative function consists of lawmaking and is

the responsibility of Congress. The governmental function implements the law and is in the hands of the Government, led by the President, followed by Ministers and Chiefs of Department. The Judiciary function administers the law; or in other words, judges and seeks to resolve conflicts between individuals or between non-State and State actors. This function is the responsibility of the courts and councils as well as the State's Attorney and Jurisdictions, and the Public Ministry (including the Citizens' Ombudsman and the Procurator General) (Cultural Division, Bank of the Republic, 2015). In addition to these three functions there are independent State bodies including the Public Ministry, consisting of the Citizens' Ombudsman and the Procurator General and the Controller General; as well as the Bank of the Republic and electoral bodies.

4 There are multiple studies about post-conflict high-value natural resource governance (see compilation by Bruch, Muffett, & Nichols, 2016; see for example Conca & Wallace, 2009; Le Billon, 2008; Le Billon & Levin, 2009; see compilation by Lujala & Rustad, 2012; Lujala, Rustad, & Le Billon, 2012). However, they rarely address community participation or deal with Latin American Indigenous peoples. The few studies that do address community participation come from extremely different contexts (see for example Chapagain & Sanio, 2012; Maconachie, 2009; Maconachie & Binns, 2007a, 2007b). The case of the Nasa of North Cauca offers fundamentally different cultural, normative, ideological, socio-political, geographical, economic and moral contexts, and territorial dynamics of mining and armed conflict so they offered an opportunity for new insights for the post-conflict natural resource governance literature. Studies about Indigenous responses to extraction in armed conflict or post-conflict Latin American contexts (see Cepek, 2012, on the Ecuador-Colombia border; see Dougherty, 2011, on Guatemala) do not explore the mining, conflict and peace nexus, where the Nasa experience offered rich opportunities in this space.

5 The place where those Canoas Resguardo members who have breached community rules and are therefore not in harmony with their community go to rehabilitate themselves.

6 All the interviews were recorded in Spanish and the translations I include here are my own.

References

Acosta, A. (2008). El Buen Vivir, una oportunidad por construir. *Debate Ecuador*, 75, 33–48.

Acosta, A. (2010). El Buen Vivir en el camino del post-desarrollo. Una lectura desde la Constitución de Montecristi. *Policy Paper*, 9(5), 1–36.

Acosta, A. (2016). Post-extractivismo: Entre el discurso y la praxis. Algunas reflexiones gruesas para la acción. *Ciencia Política*, 11(21). doi:10.15446/cp.v11n21.60297.

Acosta, A. (2017). Living well: Ideas for reinventing the future. *Third World Quarterly*, 38(12), 2600–2616. doi:10.1080/01436597.2017.1375379.

Aristotle (1948). *Politics*, trans. E. Barker. Oxford: Oxford University Press.

Armstrong, L. (2016). *The idea of a moral economy: Gerard of Siena on Usury, restitution, and prescription* (Vol. 3). Toronto: University of Toronto Press.

Bebbington, A., Abdulai, A.-G., Bebbington, D. H., Hinfelaar, M., Sanborn, C. A., Achberger, J., … Odell, S. D. (2018). *Governing extractive industries*. Oxford: Oxford University Press.

Bebbington, A., & Humphreys Bebbington, D. (2011). An Andean avatar: Post-neoliberal and neoliberal strategies for securing the unobtainable. *New Political Economy*, 16(1), 131–145.

Bruch, C., Muffett, C., & Nichols, S. S. (Eds). (2016). *Governance, natural resources and post-conflict peacebuilding*. Abingdon, Oxon: Routledge.

Cepek, M. (2012). *A future for Amazonia Randy Borman and Cofán environmental politics*. Austin: University of Texas Press.

Chapagain, B., & Sanio, T. (2012). Forest user groups and peacebuilding in Nepal. In P. Lujala & S. Rustad (Eds.), *High-value natural resources and peace building* (pp. 561–578). London: Earthscan.

Conca, K., & Wallace, J. (2009). Environment and peacebuilding in war-torn societies: Lessons from the UN environment programme's experience with postconflict assessment. *Global Governance*, 15(4), 485–504.

Consejo de Mujeres Originarias por el Buen Vivir. (2015). *Ante-proyecto deLey del Buen Vivir*. Retrieved February, 2016, from https://groups.google.com/forum/#!topic/mesacomunicacionpacifico/gxro9YfvXos

Cultural Division, Bank of the Republic. (2015). *State*. Retrieved from https://enciclopedia.banrepcultural.org/index.php/Estado

Dougherty, M. (2011). *Peasants, firms, and activists in the struggle over gold mining in Guatemala: Shifting landscapes of extraction and resistance* (Doctoral dissertation, University of Wisconsin-Madison). ProQuest Dissertations Publishing.

Escobar, A. (1995). *Encountering development: The making and unmaking of the third world*. Princeton, N.J.: Princeton University Press.

Fitz-Henry, E. (2015). Greening the petrochemical state: Between energy sovereignty and Sumak Kawsay in coastal Ecuador. *The Journal of Latin American and Caribbean Anthropology*, 20(2), 264–284.

Francis, P. (2015). Laudato si. *Vatican City: Vatican Press*, May, 24, Vatican City. Retrieved from www.vatican.va/content/francesco/en/encyclicals/documents/papa-francesco_20150524_enciclica-laudato-si.html

González Rosas, A. M. (2016). *Vivimos Porque Peleamos – Una mirada desde abajo a la resistencia indígena del Cauca, Colombia*. Ciudad de México: Memorias Subalaternas.

Gudynas, E. (2011). Buen vivir: Germinando alternativas al desarrollo. *America Latina en Movimiento*, 462, 1–20.

Hill, C. (1975). *The world turned upside down: Radical ideas during the English revolution*. Harmondsworth, Middlesex: Penguin.

Hilton, R. H. (1973). *Bond men made free: Medieval peasant movements and the English rising of 1381*: New York: Viking Press.

Hobsbawm, E. J. (1971). *Primitive rebels: Studies in archaic forms of social movement in the 19th and 20th centuries*. Manchester: Manchester University Press.

Le Billon, P. (2008). Resources for peace? Managing revenues from extractive industries in post-conflict environments. *Political Economy Research Institute Working Papers*, 143.

Le Billon, P., & Levin, E. (2009). Building peace with conflict diamonds? Merging security and development in Sierra Leone. *Development and Change*, 40(4), 693–715. doi:10.1111/j.1467-7660.2009.01568.x

Lujala, P., & Rustad, S. (Eds.). (2012). *High-value natural resources and peace building*. London: Earthscan.

Lujala, P., Rustad, S., & Le Billon, P. (2012). Building or spoiling peace? Lessons from the management of high-value natural resources. In P. Lujala & S. Rustad (Eds.), *High-value natural resources and peace building* (pp. 571–621). London: Earthscan.

Maconachie, R. (2009). Diamonds, governance and 'local' development in post-conflict Sierra Leone: Lessons for artisanal and small-scale mining in sub-Saharan Africa? *Resources Policy*, 34(1–2), 71–79. doi:10.1016/j.resourpol.2008.05.006

Maconachie, R., & Binns, T. (2007a). Beyond the resource curse? Diamond mining, development and post-conflict reconstruction in Sierra Leone. *Resources Policy*, 32 (3), 104–115. doi:10.1016/j.resourpol.2007.05.001

Maconachie, R., & Binns, T. (2007b). 'Farming miners' or 'mining farmers'?: Diamond mining and rural development in post-conflict Sierra Leone. *Journal of Rural Studies*, 23(3), 367–380. doi:10.1016/j.jrurstud.2007.01.003

Marx, K. (1992). *Capital: Volume III* (Vol. 3). London: Penguin.

Marx, K. (2000). *Capital, Volume I*, Electric Book Company, *ProQuest Ebook Central*, https://ebookcentral.proquest.com/lib/uef-ebooks/detail.action?docID=3008518.

Mayr Maldonado, J., & Olmedo Martínez, L. (2016). Indigenous peoples, natural resources, and peacebuilding in Colombia. In C. Bruch, C. Muffett & S. S. Nichols (Eds.), *Governance, natural resources and post-conflict peacebuilding* (pp. 605–626). Abingdon, Oxon: Routledge.

Owen, J. R. (2009). *A history of the moral economy. Markets, custom and the philosophy of popular entitlement*. North Melbourne: Australian Scholarly Publishing.

Polanyi, K. (2001). *The great transformation: The political and economic origins of our time*. Boston, MA: Beacon Press. https://ebookcentral.proquest.com/lib/uef-ebooks/reader.action?docID=3117969

Redacción El Tiempo. (2004, December 7). *Premio Nacional de Paz para Indigenas y Cacaoteros, El Tiempo*. Retrieved from www.eltiempo.com/archivo/documento/MAM-1525605

Reimann, C. (2004). Assessing the state-of-the-art in conflict transformation. In A. Austin, M. Fischer & N. Ropers (Eds.), *Transforming ethnopolitical conflict* (pp. 41–66). Wiesbaden: Springer.

Rodriguez-Garavito, C. (2011). Ethnicity.gov: Global governance, indigenous peoples, and the right to prior consultation in social minefields. *Indiana Journal of Global Legal Studies*, 18(1), 263–305. https://muse.jhu.edu/article/445818

Rudé, G. F. E. (1964). *The crowd in history: A study of popular disturbances in France and England, 1730–1848*. New York: Wiley.

Sandberg, J. (2015). Moral economy and normative ethics. *Journal of Global Ethics*, 11 (2), 176–187. doi:10.1080/17449626.2015.1054557.

Scott, J. C. (1977). *The moral economy of the peasant: Rebellion and subsistence in Southeast Asia*: New Haven: Yale University Press.

Scott, J. C. (1985). *Weapons of the weak: Everyday forms of peasant resistance*. New Haven and London: Yale University Press.

Siméant, J. (2015). Three bodies of moral economy: The diffusion of a concept. *Journal of Global Ethics*, 11(2), 163–175. doi:10.1080/17449626.2015.1054559.

Taussig, M. T. (2003). *Law in a lawless land: Diary of a "limpieza" in Colombia*: Chicago, Ill.: University of Chicago Press.

Thompson, E. P. (1971). The moral economy of the English crowd in the eighteenth century. *Past & Present*, 50, 76–136.

Walsh, C. (2010). Development as Buen Vivir: Institutional arrangements and (de) colonial entanglements. *Development*, 53(1), 15–21.

2 Context

Let us now look into the complex context I suggested when describing the minga or working bee against mining in the previous chapter. There are places to locate, like North Cauca, Santander de Quilichao, the Canoas Resguardo (Indigenous Reserve) or the Munchique Resguardo. There are Nasa communities and organisations, ACIN and Canoas, that are the protagonist of this book. Their identities and traditions shape how Nasa communities respond to mining. Core among them is the identity and tradition of resistance. The Nasa resistance is a response to the colonialism, dispossession and oppression that have affected their people. I will attempt to describe it here. My subsequent focus will be on mining, which many Nasa described as a vehicle of those very processes they endeavoured to resist. I shall describe mining's link not only to development discourses but to the armed conflict. By superposing these layers, I seek to illustrate the complexities of the context where the Nasa live, resist and dream. These overlapping layers of context help understand the tensions, ambiguities and conflicts the Nasa communities face when responding to mining.

Cauca and North Cauca

North Cauca is the Northern Zone of the Colombian Department of Cauca, located in the South West of Colombia (see Map 2.1). A geographically diverse region, Cauca has a share of the Andean mountains, inter-Andean valleys and Pacific coast, where it holds a section of the Choco bioregion. While Cauca has a large Pacific coast, the Northern Zone does not border the Pacific. The North Cauca area covers 13 municipalities with Santander de Quilichao, a medium-sized town, acting as the sub-regional capital (see Map 2.2). It is in Santander de Quilichao's urban area where most social organisations from North Cauca have headquarters.

Despite significant economic growth, at the time of the research, close to 60 per cent of Cauca's population remained in poverty, and the region had strongholds of illicit economies including drug production and trafficking, and illegal mining. The region's formal economy was mainly based on agriculture, livestock, forestry and fisheries (Ministry of Education of Colombia, 2012). In the years preceding my fieldwork, North Cauca's mining,

DOI: 10.4324/9781003226895-2

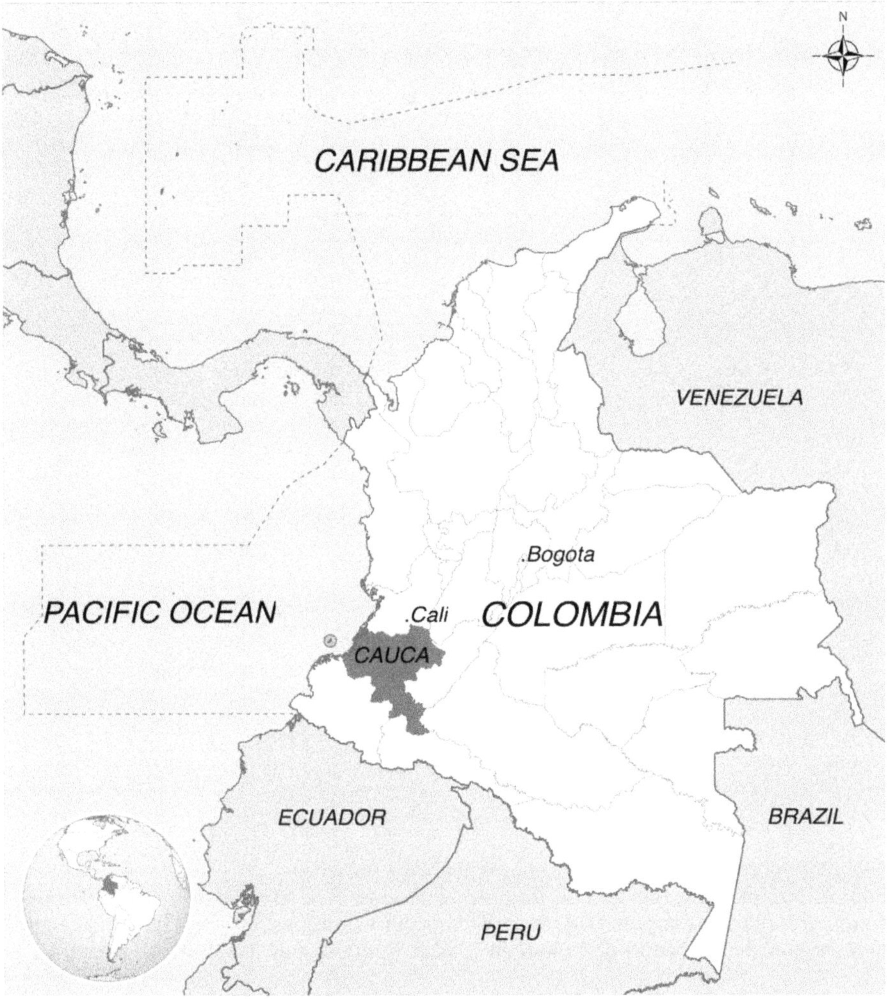

Map 2.1 Location of the Cauca Department
Source: Adapted by the author from Shadowxfox [CC BY-SA 3.0 (https://creativecommons.org/licenses/by-sa/3.0)], from Wikimedia Commons. Original file: https://upload.wikimedia.org/wikipedia/commons/a/a6/Colombia_-_Cauca.svg

construction and other industries experienced significant growth. For example, in 2013, the growth of these sectors surpassed agriculture and increased tax revenue flows to Cauca's treasury (Redacción El País, 2015). Despite this, in that same year, over 58 per cent of Cauca's population lived in poverty, including more than 28 per cent who lived in extreme poverty (National Department of Statistics, 2013). North Cauca has seen the growth of industrialised agriculture including sugar cane, corn, rice, coconut, cacao and African palm (Ministry of Education of Colombia, 2012). It also has crops of

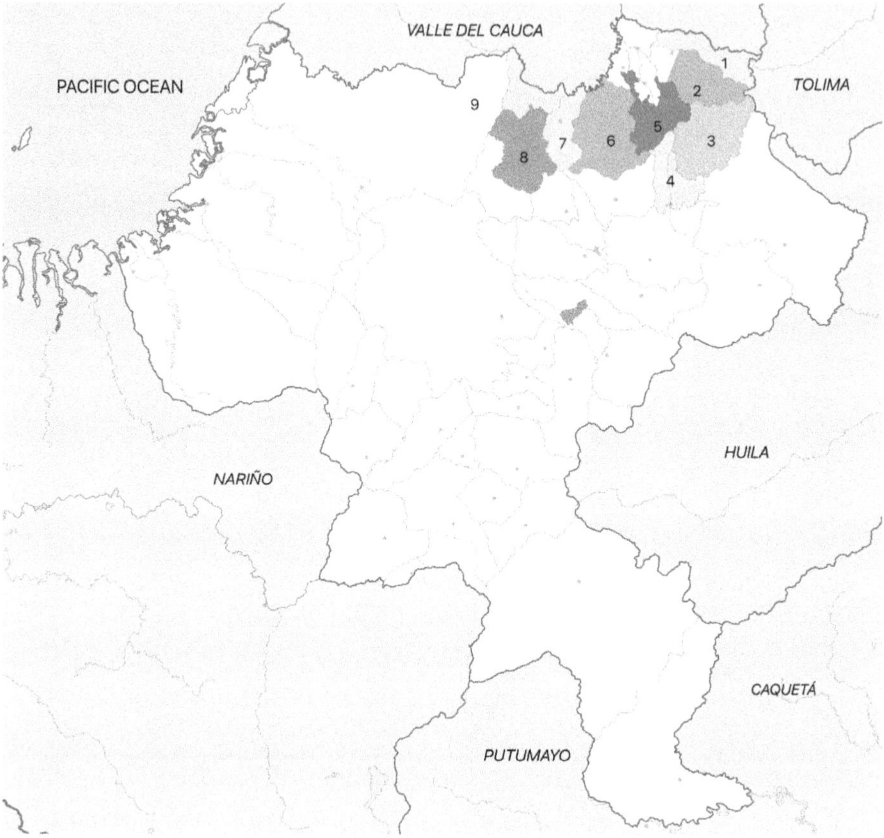

Map 2.2 North Cauca Municipalities with Nasa Presence
Source: Adapted by the author from Milenioscuro [CC BY-SA 4.0 (https://creativecommons.org/licenses/by-sa/4.0)], from Wikimedia Commons. Original file: https://commons.wikimedia.org/wiki/File:Colombia_Cauca_location_map_(%2Burban_areas).svg

Table 2.1 Legend for Map 2.2

Numbered municipalities have Nasa presence	
MUNICIPALITY	RESGUARDOS IN THE MUNICIPALITY
1. Miranda	Miranda Resguardo.
2. Corinto	Corinto Resguardo.
3. Toribío	Toribío, Tacueyó, and San Francisco Resguardos.
4. Jambaló	Jambaló Resguardo.
5. Caloto	López Adentro, Tóez Caloto, and Huellas Caloto Resguardos.
6. Santander de Quilichao	Munchique los Tigres, Canoas, Pueblo Nuevo Ceral, and Guadualito Resguardos.
7. Buenos Aires	Las Delicias, and Concepción Resguardos.
8. Suárez	Cerro Tijeras Resguardo.
9. (Northwestern) Lopez de Micay	Cerro Tijeras, and Naya Resguardos.

Source: Adapted by the author from Shadowxfox [CC BY-SA 4.0 (https://creativecommons.org/licenses/by-sa/4.0)], from Wikimedia Commons. Original file: https://upload.wikimedia.org/wikipedia/commons/a/ab/Colombia_Cauca_location_map_%28%2Blocator_map%29.svg.

traditional corn, and Indigenous communities have subsistence crops. However, monoculture expansion has brought arable land scarcity that, paired with population growth, has placed Indigenous food production under threat (Mosquera, 2015a, 2015b). Crops of illicit use are a significant livelihood source in some Cauca communities, including some Nasa communities (González Rosas, 2016; Guzmán Barney & Rodríguez Pizarro, 2015), while illegal mining, a prevalent activity in the department, has placed its waterways in jeopardy (Beltrán Rueda, 2016). This is a reason for concern because of the downstream human settlements, including Cali, Colombia's third largest city.

ACIN and Canoas

The two Nasa organisations I concentrated on were ACIN or the Association of Indigenous Cabildos of North Cauca, also known in NasaYuwe as *Cxab Wala Kiwe* that translates as Territory of the Great People, and the Nasa Cabildo of Canoas. ACIN emerged in 1994 as a means for the North Cauca Cabildos to join efforts in projects for their communities. It groups the Traditional Authorities or Cabildos of the Nasa Indigenous people of North Cauca. Canoas is a Nasa Indigenous Resguardo in the Municipality of Santander de Quilichao, with a Cabildo affiliated to ACIN.

The Nasa are a Colombian Indigenous people and as such have a suite of rights and autonomies based on international and national law and treaties. Indigenous rights have received significant recognition and undergone consolidation in recent decades through, for example, the United Nations Declaration on the Rights of Indigenous Peoples (UNDRIP) (2007) and the International Labour Organisation Convention on Indigenous and Tribal Peoples, Convention No 169 (1989) (ILO 169). At the core of collective rights platforms are these international instruments and, in the Latin American context including Colombia, national Constitutions that, since the early 1990s, began recognising political and territorial autonomies for Indigenous and tribal peoples (Political Constitution of Colombia, 1991; Van Cott, 2001). For Indigenous and tribal peoples, the international normative frameworks (The Universal Declaration of Human Rights (1948), *ILO 169, UNDRIP (2007),* World Conference Against Racism, Racial Discrimination, Xenophobia and Related Intolerance Declaration (2001)) articulate the rights to: equal treatment and non-discrimination; cultural identity and integrity; participation; preserving cultural traditions and customary institutions; being consulted in good faith with the aim of obtaining free, prior and informed consent for administrative or legislative measures that might affect them; self-determination and identifying their own development priorities; land, territory, natural resources and environment that have traditionally belonged to them or that they have acquired, respecting traditions and customary land ownership regimes; participating in the use, administration and conservation of natural resources and subsoil resources, and to being consulted should the subsoil belong to the State.

Colombia recognises Indigenous land rights, but had not implemented institutional reform to bring to reality constitutional provisions for territorial autonomy. Colombia's 1991 Constitution and ILO 169 ratification provide Indigenous and Afro-descendant communities with collective property rights. In 2005, these amounted to 30 per cent and 4 per cent of the country's land for Indigenous and Afro-descendant communities respectively (National Department of Statistics, 2005). This historic achievement resulted from social movements and Government accommodation of their demands (Velasco, 2011). However, since 1992, there have been numerous setbacks for full development of ethnic territorial authorities (Velasco, 2011). As a result, ethnic social movements contest territorial regimes that threaten their autonomy (Arbeláez-Ruiz & Viana, 2015; Velasco, 2011).

Ethnic and cultural land rights movements have a long history in Colombia with significant roots in the Cauca department. The first Indigenous land rights organisation, CRIC, was formed in Cauca in February 1971 at a meeting of 2,000 Indigenous peoples, including *campesinos*[1] and workers (Otero Bahamon, 2006; Velasco, 2011). CRIC was the response to centuries of exploitative practices by landowning elites in the Cauca department, to resistance to transformative agrarian reform, and to increasing pressures on land (Gonzáles Rosas, 2016). CRIC's political strategy was to strengthen legally recognised Indigenous Authorities and their control of territory, removing Government and church intervention (Jaramillo 2005 in Velasco 2011, p. 218). By 2011, CRIC had strong support from the vast majority of Indigenous Authorities in Cauca and represented a population of 250,000 living on 544,901 hectares of territory (González Rosas, 2016; Velasco, 2011).

CRIC became a model and platform for new regional Indigenous land rights movements and for a national movement, the National Indigenous Organisation of Colombia. This National Indigenous Organisation emerged in 1982 from the consolidation of the Indigenous movement in protest against Government attempts to legislate to contain the movement's further growth (Archila 2003 in Velasco, 2011, p. 219). Many other regional Indigenous organisations would emerge from the 1980s onwards.

ACIN is the association of Indigenous Authorities of the North Cauca Zone, part of CRIC. Its offices are in the municipality of Santander de Quilichao, the same municipality where Canoas is located. ACIN brings together 17 Indigenous Resguardos (legally recognised Indigenous territories or Reserves) and 20 Indigenous Cabildos. The Cabildos are local Indigenous Authorities made up of a Council of elected representatives. Most Cabildos have a territory (Resguardo) to administer. The Cabildos and Resguardos are located in the municipalities of Miranda, Corinto, Caloto, Santander De Quilichao, Buenos Aires, Toribío, Jambaló and López de Micay (ACIN, 2011, 2016a, 2016b, 2016c). As of the end of my fieldwork in 2017, the organisation had 180 employees and a network of 1,700 community practitioners working at the Resguardos on ACIN-coordinated programs. At the time, the population of ACIN-affiliated Resguardos and Cabildos comprised

112,313 people, just over 49 per cent of who were women (ACIN unpublished 2015 census data).

The Canoas Resguardo is a highly organised and active Nasa community, with a predominantly agricultural economy. Canoas has produced important political and organisational leadership for Nasa organisations, and has a tradition of environmental activism. Located in the rural area of the Santander de Quilichao municipality, it comprises 27 villages in the western front of Colombia's central Andean mountain chain (Canoas, 2017). The Cabildo brings together the popularly elected Resguardo Authorities. During my stay in Canoas, 35 people had positions as Cabildo Authorities. They were responsible for all aspects of life in the Resguardo, from legal investigations to the administration of justice, to leading the education and health programs. The most important Cabildo participation space is the Assembly, where 2,000 to 3,000 of the more than 9,300 community members take part, three times a year. Canoas leaders are active in cultural and environmental projects aimed in particular at young people. An example is Cxayuce Arte y Cultura, which offers political and dance education to Canoas youth, who regularly perform Andean dances like Caporal at local and regional events. At the time of my fieldwork, Canoas was home to less than 8.3 per cent of the Nasa population of North Cauca. Forty-nine per cent of the population was female (Canoas, 2017). Reputed for its high agricultural output, the Resguardo produced coffee, fruits and starchy foods and, at the time, had been successful in eradicating crops used for the drug trade. It had struggled with illegal miners entering the territory, but during my visit, the problem had been controlled to some extent and had migrated to the nearby Resguardo of Munchique, where we attended the minga. Nevertheless, Canoas faced difficulties providing employment for its population, which was predominantly young according to unpublished ACIN census data from 2015. This was a difficulty that motivated youth focused efforts such as Cxayuce's.

The Nasa are known nationally and internationally for their level of organisation and their peacebuilding and resistance work during the Colombian armed conflict (González Rosas, 2016; Rappaport, 2000). They have been pioneers in the Colombian Indigenous rights movement, often leading Indigenous proposals and struggles as active participants of CRIC. Amid the parallel and mutually reinforcing processes of colonialism, dispossession and oppression; armed conflict and violence; and mining enterprise in Indigenous territories, the Nasa have consolidated a resistance identity. Rappaport (2000) observed that a Nasa resistance ideology had formed during colonial times and continues to be updated. She states that the Nasa cosmogony[2] is strongly linked to resistance tactics (Rappaport, 2000). The resistance has taken a range of configurations. Until the seventeenth century, they concentrated on fighting the Spanish. After that the strategy shifted to accommodation to avoid extermination. During the independence period, in the early nineteenth century, the expression of resistance was either patriot or royalist positioning, according to interpretations of what side would best protect Indigenous interests (Proyecto Nasa, n.d.). In recent history, resistance has encompassed a wide range of strategies including establishing organisations, campaigning for Indigenous rights and staging large protests; participating in an armed self-defence movement, or declaring territories free of weapons and armed conflict; and structured proposals for significant national legal and policy reform. As Nasa leaders have historically become more knowledgeable of the rights frameworks and laws, legalism became a central strategy in Indigenous resistance (González Rosas, 2016; Rappaport, 2000). In contemporary Nasa Indigenous organisations, resistance is expressed not only in actions such as legal battles or when the communities declare an area a territory of peace, but in the Nasa's positioning in conversations, their way of operating and communicating. Face-to-face and written communications routinely start and conclude with, at a minimum, an explicit mention of the resistance struggle, if not a narrative about the current resistance work. Resistance is so central that Rappaport concluded it is part of the habitus (following Bourdieu, 1977) of the Nasa community (Rappaport, 2000).

Colonialism, Dispossession and Oppression

The Nasa are both survivors of, and active participants in, Colombia's tumultuous history, a history marked by successive periods of long-lasting war, violence and oppression. Numerous Indigenous cultures, including the Muisca, Tairona, Chibcha, Caribe, Pijao and Quimbaya, lived in Colombia before the Spanish colonisation in the fifteenth century. They left a diverse legacy of gold work, ceramic and adaptive agriculture. The Spanish invasion and rule was primarily a large-scale extractive enterprise, conducted in a violent and exploitative fashion (Bebbington & Bury, 2013). The Spanish invaders decimated the Indigenous population and used African people as slaves. The independence in 1810 brought its own problems including capital flight and difficulties in establishing an autonomous government (LaRosa & Mejía, 2012). The legacy of the stratified, exploitative social structure and the oppression of the Spanish rule still lives today. Colombia's income inequality is amongst the highest in the world and it presents 'stark' inequality among its diverse regions (Organization for Economic Co-operation and Development, 2015). Descendants of the *hacendados*[3] occupy the place where the Spanish rulers once were (Ospina, 2013). Land concentration reflects this inequitable pattern and Colombia has been identified as the most unequal country in Latin America for land distribution (Guereña, 2016).[4] These have motivated social movements to seek economic equality and land reform (Pizarro Leongómez, 2015). The response to social protest has followed a consistent pattern of violent repression (Moncayo Cruz, 2015; Vega, 2015).

Cauca has a strong Spanish occupation heritage with an economic history characterised by value extraction through displacement, exploitation, servitude and slavery. Cauca's capital, Popayán, was an important regional government centre during the colonial period (Gamarra Vergara, 2007). Popayán houses one of the oldest Colombian universities (Universidad del Cauca) and has a deeply engrained Catholic tradition. Cauca has high concentrations of Indigenous populations native to the region, and Afro-descendant populations who are descendants of people forced into slavery (National Department of Statistics, 2007b). Descendants of the Spanish colonial time families of Cauca – hacendados – are today part of the ruling political class of Colombia, with families counting presidents, ministers and senators in their genealogies. Dispossessed of their lands and forced to surrender their work unpaid to hacienda owners, Indigenous populations languished as a result of the harsh treatment they received from the colonisers (Finji & Rojas, 1985). During the colonial period, the crown 'protected' Indigenous people and concentrated them on dedicated land or Reserves (Resguardos) to ensure they were available for ongoing servitude conditions (Finji & Rojas, 1985; Molano, 2015b). Today Cauca has one of the most unequal land distributions in Colombia (Organization for Economic Co-operation and Development, 2015). Land concentration in North Cauca is reflective of this broader departmental dynamic (see Duarte, 2015).

Cauca is home to some of the strongest Indigenous and cultural social movements in Colombia (Otero Bahamon, 2006). Their leaders have been targeted with selective violence (González Rosas, 2016). Significant victories in the land rights front made Indigenous leaders in Cauca targets for repression and 50 Indigenous leaders were subjected to forced disappearance[5] during 1971 and 1972 (Otero Bahamon, 2006). During the 1970s, paid armed groups called 'the birds' (*los pájaros*) conducted murder and intimidation campaigns against the leaders of the land recovery Indigenous movement (González Rosas, 2016). As a form of Indigenous defence, the Indigenous armed movement Quintín Lame (MAQL) emerged in 1983. A few years later, lack of community support for the armed group led the Quintín Lame to demobilise (Guzmán Barney & Rodríguez Pizarro, 2015). In a peace agreement with the Betancourt Government in 1986, Quintín Lame achieved land title for over 12 million hectares and the creation of 151 new Indigenous Resguardos (Otero Bahamon, 2006). This remains a landmark achievement for the Indigenous movement of the region after its attempts to recover land illegally acquired and extracted from Indigenous Reserves. In 1991, armed men paid by a drug dealing syndicate, in collaboration with the Santander de Quilichao police, massacred 20 Indigenous people at the El Nilo Hacienda, while the Indigenous people were on a campaign to have the Government purchase the Hacienda and give the title to them (Guzmán Barney & Rodríguez Pizarro, 2015). A settlement was later reached where the Government was ordered to remediate the Nasa community with the purchase of land that at the time of the study had not been fully delivered.

During the time I was doing this research, violence against Cauca leaders continued and Nasa leaders were not exempt. Criminal gangs widely advertised murder threats against Indigenous leaders and defenders of human rights and carried them out. The Colombian Government engaged in violent repression of protests in Cauca and elsewhere in Colombia, through its specialist anti-protest police force known as the ESMAD (mobile anti-riot squadron). The pattern of violence against Indigenous and social leaders in general continued and placed Cauca leaders at particular risk. In July 2018, the Colombian Ombudsman reported that 70 per cent of the 311 social leaders murdered in Colombia since 2016 did community work in rural areas and were murdered because they opposed two activities: drug trafficking and illegal mining. Seventy-eight of these leaders were from Cauca (Cárdenas, 2018). By December 2018, the number of Cauca leaders murdered had risen to 103 (Colombian Ombudsman, 2019).

Conflict and Peace

Colombia has been immersed in internal armed conflict for more than half a century. It can be argued that no living Colombian has known a country in peace. The conflict has had harrowing effects on individuals, families and the entire nation. It has been complex, prolonged, geographically dispersed and plagued with atrocities (Pizarro Leongómez, 2015). Most accounts place the rural economy, in particular land rights, social inequality and institutional weakness, as central triggering factors, causes or at least favouring conditions for the conflict (Fajardo, 2015; Molano, 2015a; Pizarro Leongómez, 2015). The history of inequality, economic exclusion and violent oppression of the campesino movement in particular is prominent in narratives that explain the conflict (see Molano, 2015a). There is consensus in historical accounts that there have been a series of 'fault lines' in the construction of the Colombian nation and that the prolonged postponement of crucial social reform, in particular land reform, has fuelled the conflict (Pizarro Leongómez, 2015).

Some of the historic narratives speak of the pervasive links and vicious cycles of repression, displacement, land concentration and depreciation, increased dissatisfaction and rebel activity. Triggering the cycle were repressive violence against campesinos and their organisations in the rural areas of Colombia, together with violent displacement of campesino, and rural Indigenous and Afro-descendant communities. These would lead to land depreciation, which would support the concentration in private hands of the land previously owned, or worked, by now displaced communities. Community and rebel group dissatisfaction with the rural economic order would grow. The result was the emergence of, or increased activity by, armed rebel organisations (Molano, 2015a).

Other economic factors would compound these issues. For example, lucrative extortion activities aimed at the cocaine and petroleum businesses allowed Colombia's rebel groups such as the FARC and the National Liberation Army (ELN) to grow their ranks to thousands of militants (Molano, 2015a). Violence and the lack of economic opportunities in rural areas facilitated recruitment into these organisations or the drug economy that financed them. In a vicious cycle, the drug economy became an obstacle to economic opportunities in rural areas.

Class, an important fault line in the Colombian conflict, has complex interactions with multi-ethnic dynamics that make it necessary to study ethnic perspectives in particular as they relate to natural resources. It has been argued that the conflict originated around the violent disputes for control of the State and of land (Molano, 2015a) where class was a determinant of this access. Class was a fault line in the conflict's watershed periods such as the State repression of organised socialist campesino (peasant) leagues and communist cells of the first half of the twentieth century (Molano, 2015a); and the insurgency and counter insurgency of the second half of the twentieth century, when Marxist inspired guerrillas confronted capitalism and met the counter insurgency of legal and illegal elites (Moncayo Cruz, 2015). The campesino leagues represented the rural workers and included Indigenous members, as did the guerrillas. Nasa leaders such as Quintín Lame had been central to the land rights movement in the first half of the twentieth century, while also advocating a strong Nasa Indigenous identity (Rappaport, 2000; Molano, 2015a). Although the Colombian conflict is not an ethnic conflict, the observations of Kearney and Varese (2008, p. 209) about Latin American societies that "*a complex class structure [...] cuts across and permeates the multi-ethnic configuration*" apply to the Colombian case. The Indigenous differentiation within rural communities continues to strengthen today in Colombia, paralleling the trend of 're-indigenisation' (Jackson & Warren, 2005) of Latin American communities once classed as peasants (Dougherty, 2011) that have gained recognition of their land rights.

The number of actors involved and Colombia's geographic complexity complicated and prolonged the conflict. The multiple actors include a fractured and poorly coordinated set of national, regional and local level State institutions; multiple guerrilla movements with diverse political and military projects (M-19[6] and MAQL – demobilised in the early 1990s, FARC – in the process of demobilising at the time of the research, ELN – active and in peace negotiations at the time of the research with negotiations later halted, EPL[7] – disbanded); and highly atomised paramilitary groups (Pizarro Leongómez, 2015, p. 45). Colombia has a complex geography and the conflict has been geographically disperse. These contributed to making Colombia's armed conflict the longest in the Latin American region (Pizarro Leongómez, 2015).

The Colombian conflict involved many atrocities that numbers cannot fully describe. Between 1985 and 2000, for each combatant dead there were an estimated 80 civilians killed, and this went as high as 380 in the following years (Giraldo, 2015; Pizarro Leongómez, 2015). More than 7,358,248 people were forcibly displaced, 995,393 were murdered, at least 169,201 were forcibly disappeared, 36,578 were kidnapped and 10,787 were tortured (Centro Nacional de Memoria Histórica, 2018).[8] The total number of victims is estimated at over eight million (Centro Nacional de Memoria Histórica, 2018). Indigenous people and Afro-descendants have been seriously and disproportionately affected by violence and forced displacement (Constitutional Court of Colombia, 2009a, 2009b; National Department of Statistics, 2007a, 2007b; Urrea & Viáfara, 2007;

Comisión de la Verdad, 2022). The consequences of the conflict on Indigenous people are not only measured in statistics but in the cultural effects of youth recruitment, forced displacement or confinement, restrictions to everyday activities and free movement, and the use of Indigenous communities as human shields (Comisión de la Verdad, 2022).

North Cauca witnessed some of the most intense armed conflict in recent history. FARC and ELN presence was traditionally strong in Cauca. FARC undertook aggressive recruitment campaigns in North Cauca including child recruitment (Fundación Ideas para la Paz, 2013). During the fieldwork, participants shared with sadness their first-hand accounts of youth and child recruitment experiences. Where FARC has traditionally taken its militants away from home, Nasa members of FARC have remained in their home territories, leading to a unique configuration of local armed conflict dynamics.[9] In the early 2000s, paramilitary groups arrived and brought the violence to new heights (see Taussig, 2003). During the 1990s and 2000s, combats between the army and guerrillas, as well as ambushes, bombings, landmines, town sieges, kidnappings and selective killings were common, if not daily, occurrences in several North Cauca municipalities with large Nasa populations such as Toribío, Caloto and Corinto. Adding to this, paramilitaries murdered and harassed several Indigenous leaders and held open combat with FARC in Indigenous territories.

A study by Guzmán Barney and Rodríguez Pizarro (2015) identified three key junctures defining significant periods in the armed conflict in North Cauca. First, from 1990 to 1992, the M-19 and Quintin Lame guerrillas demobilised. These were partial processes that simultaneously saw the expansion of FARC militias in the mountain areas and of drug trafficking in the flat lands. Both came with extreme forms of violence and crime. Second, from 1999 to 2004, the paramilitaries arrived in North Cauca. Their intention was to drive away FARC. In the process they "*oppressed civil society organisations, silenced State institutions and delegitimised Government armed forces*" (Barney & Pizarro, 2015, p. 102). Third, from 2008 to 2010, there were combats, ambushes and sieges in Caloto, Corinto, Caldono, Toribío, Jambaló and Santander de Quilichao, together with growth in the drug trade, including laboratories on the flat lands and crops in the mountain areas, and the proliferation of unidentified violent groups. Although the paramilitaries officially demobilised in 2006, false demobilisations and dissidence plagued the process, and from them emerged new so-called criminal bands pursuing similar agendas and undertaking similar forms of violence.

The confrontation between guerrillas, paramilitaries, drug trafficking gangs and criminal bands for control of the Nasa territories and people had a marked effect in Nasa communities, which a report by the Observatory of Human Rights and International Humanitarian Law of the Vice-president's Office explains thus (2009). The region is strategically located for its access to the Pacific region; and its value as a retreat area for FARC, in particular the highlands, where FARC had also promoted the proliferation of crops for illicit use (marihuana, coca and poppy), including in the predominantly Indigenous municipalities of Toribío and Jambaló, and in Corinto (with a high Indigenous population), all of which are consistently among the municipalities most affected by illicit crops (United Nations Office on Drugs and Crime & Government of Colombia, 2016). The presence of FARC made the region an area of territorial contention with the paramilitaries. The interest of the drug trafficking gangs from the neighbouring Valle del Cauca department was in the lowlands such as Santander de Quilichao because of their proximity to drug processing facilities. These actors "*clashed with the collective interests of the [...] Nasa people. Through intimidation and armed coercion these groups sought to penetrate the communities, expropriate their territories, disintegrate them culturally and involve them in the armed confrontation*" (Observatory of Human Rights and International Humanitarian Law of the Vice-president's Office, 2009, p. 9). The Nasa community met these with resistance that turned its leaders into targets. However, the Nasa did not let up. I heard repeatedly during the fieldwork that the determination was to not allow any armed actor to displace the communities from their territories but instead to confront them in large numbers, to precipitate their exit. Likewise, it was often explained to me that this was no easy task because the crops of illicit use, their economy of easy profiteering, and the networks of violence that the armed actors created inside Nasa communities had turned into obstacles.

This book concentrates on the post-agreement transition period, a crucial moment for Cauca considering its history of violent conflict. Having suffered such intense and prolonged violence, it should not come as a surprise that, when asked in a plebiscite in late 2016, Cauca voted overwhelmingly in favour of the *Agreement to end the armed conflict and build long lasting and stable peace* between the Government of Colombia and the Guerrillas of FARC (the Peace Agreement). It was during that transition period, including the late stages of agreement negotiation, signing, plebiscite, renegotiation and early stages of implementation, that I collected the data for this work. Let us look briefly into what the peace process entailed.

The year 2016 saw the conclusion of Colombian Government–FARC EP peace negotiations, a process that began publicly in 2012. In August 2016, the parties reached an agreement and it was formally signed in a large-scale ceremony in Cartagena, Colombia in September 2016. However, this was not the first peace negotiation process between the Colombian Government and FARC; previous attempts by three different Colombian Presidents failed in the 1980s, 1990s (with a coalition of guerrilla groups) and early 2000s. So, one of the most significant obstacles of the peace process was community distrust. This was evident in the results of the Peace Agreement plebiscite held in October 2016. Through the plebiscite, Colombian president Juan Manuel Santos sought citizen endorsement for the Peace Agreement ahead of implementation. The vote against the Peace Agreement won by a narrow majority, demonstrating just how divided, apathetic or otherwise constrained Colombian society was when it came to taking decisive action. Only 37.4 per cent of voters participated, 50.2 per cent voted against and 49.8 per cent voted for (Registraduría Nacional del Estado Civil, 2016). After the NO victory, core sections of the agreement were renegotiated, in particular sections on rural reform, precisely the area of the agreement that was of most urgency based on analyses of the Colombian conflict. After that, Congress unanimously endorsed the Agreement, without convening a second plebiscite. Insiders of the NO campaign later acknowledged that the campaign was based on propaganda designed to cause anger and division (Redacción Política, 2017) and the Colombian Courts have issued decisions confirming this was the case (Consejo de Estado, 2016).

In the peace process, mining was not part of the agenda for negotiation; instead was one of many themes the Santos Government expected to address through a 'territorial peace' approach during agreement implementation (Delegates of the Government of Colombia and FARC, 2012, 2016; Jaramillo, n.d.). The parties to the Peace Agreement identified six areas for negotiation that make up the final endorsed agreement: agrarian reform, victims, end of the conflict, drug trafficking, political participation, and agreement verification and citizen endorsement. Drug trafficking, an important source of financing for the armed conflict, had a dedicated chapter in the Agreement, but the same approach was not applied to mining, despite playing a comparable financing role. However, the negotiation logic of the Peace Agreement included the concept of 'territorial peace' where many specific questions or themes, like mining, were to be addressed during the implementation stage in each territory using the principles and governance mechanisms specified in the Peace Agreement.

The idea of 'territorial peace' aimed to build peace in Colombia's regions and form a basis for Agreement implementation as follows. The premise was that the Peace Agreement should transform the Colombian territories, in particular rural areas, so that people can have guarantees for their social, economic, civil, cultural, environmental and political rights. This transformation involves institutional strengthening and citizen participation to build a shared vision of each territory. State institutions, civil society organisations, communities, the private sector, the church, academia and all citizens must build territorial peace together. The practical means to build peace in the territories shall be social dialogues in the regions, territorial compacts amongst territorial actors and dialogue between the National Government and the regions (Office of the High Commissioner for Peace, n.d.).

The idea of territorial peace was contingent on participation and dialogue, and by implication it was contingent on understanding the ways communities living in the crossroads of mining and armed conflict think about both. The Agreement provided a set of principles and priorities for the implementation phase that can guide some aspects of dialogue. For example, territorial focus is a principle, the territories most affected by the conflict are a priority and there is a stated aspiration to implement holistic rural reform. Furthermore, after a drawn-out campaigning process, representatives of Indigenous and

Afro-descendant peoples accessed a space in the Agreement negotiation process. This resulted in an Ethnic Chapter that provided principles to protect the rights of the Ethnic Peoples and to address the effects of the conflict on their communities and territories (Delegates of the Government of Colombia and FARC, 2016). Mining has cross-cutting relevance to the principles in the Ethnic Chapter. Some study participants believed that it could be in seeking to apply these principles that the opportunities to deal with mining-conflict interactions could become more apparent.

Mining, Piñatas, Locomotives and Guns

Colombia has had difficulty in organising its mining sector and dealing with its deleterious effects, including its negative interactions with social and armed conflict dynamics. Over 70 per cent of Colombia's mining activity is informal and there are only a handful of large-scale mining operations (Echavarría, 2014). Nevertheless, successive Colombian Governments positioned large-scale mining as a force of development (Contraloría General de la República, 2014; Government of Colombia, 2010; Ministry of Mines and Energy, 2006). A good proportion of socio-environmental conflicts in Colombia are linked to mining and there is a statistical correlation between the extractive platform of previous Governments and the incidence of social conflict (Pérez Rincón, 2014). With regards to mining's interaction with the armed conflict, the scenario is not more encouraging. The former Colombian Police Director and Colombian Vice-president General Naranjo stated that gold mining is worse than the drug trade (Naranjo in Rettberg & Ortiz-Riomalo, 2016). Rettberg and Ortiz-Riomalo (2016) found that, although gold is a legal resource, its links to criminality, violence and the drug trade might make its effects on armed conflict as pervasive as those of illegal resources such as drugs. Links have also been found between mining of other minerals and the armed conflict (Centro Nacional de Memoria Histórica, 2016; CINEP, 2012; Massé & Camargo, 2012; Moor & Sandt, 2014; PBI Colombia, 2011).

Why talk about piñatas when talking about mining in Colombia? The piñata is a feature of Latin American children's parties, a moment when the children are literally showered with goodies because the piñata, a parcel filled with toys and confectionery that is suspended in the air, has been broken with a stick, to release the valuable cargo for the children to take. When the piñata breaks the children do their best to take a good share of the bounty amid the ensuing chaos. In Colombia, there is a period in the concession granting record that Colombians now know as the 'Mining Piñata' (*La Piñata Minera*) because of the disorganised and indiscriminate granting of mining concessions. In the early 2000s, the Colombian Government initiated a campaign to increase mining investment in Colombia. In that respect, Rettberg and Ortiz-Riomalo (2016) state the institutional framework was successful in attracting investment but had environmental protection and technical flaws. The Uribe Governments embarked on a rapid concession-granting enterprise to attract such investment (Negrete, 2013). Former State Controller[10] Sandra Morelli has described it as an excessive, opportunistic and disorganised concession-granting process (Garay, 2014; Morelli, 2014). Nowadays, it is widely recognised as such, hence the name 'piñata'. As a result of that disorderly approach, titleholders were not chosen based on capability (Garay, 2013). The extent of concession granting was such during the mining piñata that it created a significant secondary market for mining titles.[11]

After the mining piñata came the Santos Governments (2010–2018), and with them what came to be known as the Mining Locomotive (*La Locomotora Minera*). The Santos Government saw mining as an engine for development, and identified it as one of five "*development locomotives*" (*locomotoras del desarrollo*) that would pull the Colombian economy into the future (Government of Colombia, 2010).[12] The First Santos Government National Development Plan aimed to increase mining investment as a way to raise Government revenues. The goals included ramping up gold production by 47 per cent, coal production by 69 per cent and geological knowledge by 30 per cent, as well as shortening the mining titling period from 17 to three months. Institutional reform, formalisation, increased productivity and better information systems would serve to achieve the goals (Government of Colombia, 2010).

However, after the institutional reform, Colombia's highest fiscal control body, the General Controller, commissioned a multi-year academic study that found significant weaknesses in Government mining sector institutions, and an overall lack of capacity to regulate a highly complex system (Morelli, 2014). The General Controller's study described a disorganised, poorly regulated and highly informal sector, with limited integration into local economies and concerning deleterious effects on ecosystems and on community rights (Garay, 2013, 2014; Morelli, 2014). This presented a discouraging scenario for a post-agreement transition. Some of the concerns the study highlighted were as follows. Large-scale mining contracts minimised room for operator responsibility and Government benefit. Asymmetries of information and technical capacity between large-scale mining companies and Government institutions placed the latter at great disadvantage. The size of the artisanal and small-scale mining sector was well beyond Government formalisation capacity. Illegal and criminal mining were intimidating communities or, in some cases, maintained convivial relationship with local authorities. There was illegal exploitation of coltan, gold, uranium and tungsten in areas of environmental and cultural significance such as the Amazon. The environmental licensing framework was insufficient. State institutions were unable to organise the mining system to prioritise public over private interests. Surface and subsoil rights were confused, resulting in a pattern of community displacement and loss of agricultural land in favour of subsoil rights. There was no evidence of large-scale mining integration with regional economies or of socio-economic improvements from royalty flows. Vulnerability analyses, including from a rights perspective, were absent, even where the groups were susceptible to affirmative action on their rights. Although the rights of women, ethnic groups and campesino communities are of a higher legal hierarchy than licencing rights, the hierarchy was not observed and instead legal protection and recognition were lacking. Furthermore, few royalties flowed to victims of armed conflict and ethnic groups (Morelli, 2014; Garay, 2014).

This suggests that neither the mining piñata nor the mining locomotive delivered on their promise of prosperity. Former Controller Morelli (2014) argued that in order to implement reforms that could address the resulting conflicts, it was necessary to study those conflicts and the dynamics of social mobilisation, including from a human rights perspective. In these reforms, she argued, mining policy should be placed as central to historic opportunities to address conflict, such as the peace process. This brings us to the third part of the mining story I am revisiting here, the 'guns'. Former Controller Morelli's call for studies into mining-conflict interactions was more than justified when we consider that: i) the presence of paramilitaries in mining regions resulted in massacres, murder and violations of human rights (CINEP, 2012; Moor & Sandt, 2014; Weitzner, 2012); ii) this violence, together with competition for resources, put ethnic livelihoods at risk (CINEP, 2012); iii) Colombian judges have documented the role of mining in the armed conflict (see Constitutional Court of Colombia, 2009a, 2009b; see Supreme Tribunal of Antioquia, 2014; see Supreme Tribunal of Bogotá, 2014; see Arbeláez-Ruiz and Viana, 2015, p. 15) and iv) social leaders who oppose illegal mining are at increased risk of intimidation and violence (Cárdenas, 2018).

Scrutiny of the causes of Colombia's internal armed conflict, historic accounts of the conflict and examinations of the finance portfolios of armed groups place mineral and petroleum wealth as a central source of finance (Massé & Camargo, 2013; Molano, 2015a; Rettberg & Ortiz-Riomalo, 2016). During the 1960s, the Cuban Revolution began a wave of Marxist inspired insurgent movements in Latin America that reinvigorated the guerrilla war in Colombia. However, guerrilla groups, such as FARC and ELN, weakened during the 1960s and 1970s, their numbers dropping significantly (Molano, 2015a). It was during the 1980s that these guerrillas were able to tap extractive industry wealth by extorting oil companies. These moneys aided guerrilla re-expansion (Molano, 2015a; Verdad Abierta, 2015b). At the time I started fieldwork for this book, illegal mining had been identified as a growing element in illegal armed groups finances, its importance competing with the drug trade (Rettberg & Ortiz-Riomalo, 2016). Both guerrillas and sections of former paramilitary groups (known as criminal bands) profited from the illegal mining business that they protect through the use of violence (Colombian Ombudsman, 2010). Government armed forces had been accused of complicity, with a common complaint that "*everyone, but the police/army, sees the large excavators of illegal miners coming in*" (Massé & Camargo, 2012, p. 41). Extortion practices were ongoing, not only with oil, but also with mining companies and smaller-scale operators. This phenomenon persisted despite multi-stakeholder efforts to control it, and some attribute its pervasiveness to the frequent use of contractors with limited operator oversight in the mining sector (Massé & Camargo, 2012).

Interactions or involvement of guerrilla and paramilitaries with oil and mining extraction contributed to environmental degradation, either by controlling mineral extraction and applying poor extraction practices, or through sabotage of extraction infrastructure. Criminal/paramilitary-related groups such as the criminal bands have profited from the expansion of highly polluting criminal mining activity (Colombian Ombudsman, 2010). From the 1980s until the present day, guerrilla movements, in particular the ELN, have on several occasions bombed oil pipelines, causing extreme environmental damage (Molano, 2015a).

Colombian armed-conflict actors often gravitate towards extractive activities and enclose the surrounding communities in a veil of violence. For example, a common saying in Colombia is that 'each oilfield has its own army battalion and guerrilla front'. It is well known that the Colombian Government has actively provided security for mining operations (Massé & Camargo, 2012). As a result, the host communities can end up in the crossfire between Government forces and other armed actors, or be at the receiving end of other forms of violence that those defending unwanted mining activity use to assert their presence (Massé and Camargo, 2012). Criminal bands use violence against women to silence community opposition to criminal mining in regions like Cauca (Machado, 2015). Unfortunately, rural dwellers, including Indigenous, Afro-descendant and campesino communities, have been disproportionally affected by the confluence of this violence with other deleterious effects of mining activity (CINEP, 2012). Social leaders who oppose illegal mining (and drug trafficking) are at high risk of violent death (Cárdenas, 2018). There are also documented examples of links between illegal paramilitary groups and large-scale multinational mining companies (Moor & Sandt, 2014).

Several Colombian courts and research pieces have identified the links between illegal armed actors, the armed conflict and the mining sector. The Constitutional Court documented links between mining and conflict, identifying mining as an underlying factor in the country's armed conflict and in ethnic communities' displacement (Constitutional Court of Colombia, Decisions 004 and 005 of 2009, following from Decision T-025 of 2004). Special jurisdiction courts have raised concerns over the involvement of some mining companies in human rights violations (Supreme Tribunal of Antioquia, 2014), including through links to illegal armed actors and by providing them with financial support (Supreme Tribunal of Bogotá, 2014).

Cauca has had its share of this story of piñatas, locomotives and guns. Analyses by ACIN and others indicated that significant areas of Indigenous Reserves had been granted as part of a number of mining concessions (Montaño, 2014). There were no large-scale mining operations in Cauca, although some global mining companies hold or have held concessions (see p.

42). Illegal mining in Cauca is of such a scale that the States Attorney's Office dedicated a unit to investigate it. Precise estimations of the scale of illegal mining in Cauca are hard to arrive at because it is a clandestine activity that often has links with violent, criminal actors. The devastation that illegal mining has caused in North Cauca motivated Afro-descendant and Indigenous communities to mobilise to demand a National Government response (Ebus, 2014).

The first two decades of the twenty-first century saw the granting of numerous mining concessions in the Cauca department. Cauca had large extensions of its territory titled (14.5 per cent) or subject to application and this included areas with agricultural or environmental vocation (Montaño, 2014). The legislation around priority rights for Indigenous communities was such that some communities felt forced to do mining against their will to prevent external miner entry (Montaño, 2014; Viana, 2015). In 2014, ACIN did a study bringing together Government data and community research that described the situation in Cauca as follows (see Montaño, 2014). The study indicated that in Cauca, a department of 3,101,500 hectares, mining titles and mining legalisations covered 450,000 hectares. Cauca was second after the Antioquia department both in absolute surface area and in number of concessions allocated to mining. More than 250 mining titles[13] were valid, largely for construction materials (149 or 58 per cent of titles), followed by gold and associated minerals (68 or 26.6 per cent of titles). There were titles for sulphur, limestone, coal, nickel, precious metals and precious stones. Indigenous Resguardos or Cabildos held six of these titles (927 hectares) mostly for construction materials, sulphur and precious metal mining. More than 21 per cent of titled areas were agricultural lands, according to Government inventories, and 65 per cent were natural forests. Several large-scale mining companies had titles in Cauca.[14] A total of 771,894.82 hectares were subject to application mostly for mining gold, copper and precious metals. Seventeen out of 394 applications were from Indigenous Cabildos or Resguardos. These applications were 'priority rights' motivated. Priority rights allow Indigenous land titleholders priority over other mining title applicants. Communities had applied for titles to protect their territories from external interests. This required them to mine the land to maintain title (Montaño, 2014).

The same study by ACIN (Montaño, 2014) indicated that in North Cauca, existing titles and applications were largely for actors external to the Indigenous community, but there had also been community title legalisations and attempts to protect the territory through Indigenous priority rights that require communities to practice mining (Montaño, 2014). I summarise highlights from the ACIN study as follows (see Montaño, 2014). The three municipalities with most mining titles or requests for titles were Suárez, Santander de Quilichao and Jambaló. There were 66 mining titles. Thirty-two mining titles had been legalised, mostly for construction materials mining, but also for gold (eight) and coal (one). Two of the gold legalisations were for the Nasa Resguardo of Las Delicias. There were 59 mining title applications, 12 of which came from Resguardos. Among the corporate titleholders at the time of the ACIN study were Anglo Gold Ashanti Colombia with three titles, Anglo American Exploration with two, and Condor Precious Metals with one. Together these titles overlapped with 15 villages in the Concepcion, Guadalito and Canoas Resguardo areas of influence, and with three villages in the Corinto Resguardo area of influence. Corporate title applications included 12 requests each for Anglo American Exploration[15] and Anglo Gold Ashanti Colombia, one for Four Points Mining, and one for South American Exploration and Finance. All of them were for gold. Anglo Gold Ashanti had five applications suspended in Suarez, it had relinquished another four and two more were pending at the time of the ACIN study. Anglo American had eight applications pending, two rejected or pending appeal, one suspended in Suarez and one overlapping a natural resource management area at the time of the ACIN study (Montaño, 2014).

The growth in the mining titled area paralleled the growth in monoculture, Indigenous agricultural land scarcity, increasing efforts by Afro-descendant communities to obtain collective titles for their ancestral lands and campesino movement expectations for new campesino Reserves (territories) in Cauca (Verdad Abierta, 2014, 2015a). Territorial conflicts already existed amongst communities and with the monoculture industry. In addition, communities' perspectives about mining were not always aligned. Existing land pressures appeared to be growing with the increase in mining activity.

Discussion

In Colombia, the response to the mining-armed conflict nexus has been at best fragmented and the peace agreement negotiation did not provide a careful response to mining dynamics. The lack of a considered response to questions of mineral resource extraction is a frequent gap in peace settlements globally (Blundell & Harwell, 2016). In the Colombian context, leaving these matters unaddressed created uncertainty for three main reasons. First, because mining has been identified as a complicating factor in the Colombian armed conflict (Valencia & Riaño, 2017). Second, because it was earmarked as a funding and development source for the post-agreement transition (Ministry of Mines and Energy, 2016). Third, because of mining's interaction with fault lines of the conflict such as access to land. Considering: i) the rapid expansion of environmentally and socially damaging forms of mining (Valencia and Riaño, 2017), ii) previous Government efforts to attract international investment in mining (Government of Colombia, 2010, 2013), and iii) the established role of illegal mining in the finances of illegal armed groups (Rettberg & Ortiz-Riomalo, 2016), questions as to the role of mining in the post-agreement transition need urgent answers. Those answers must respond to the dynamics of the regions and communities that have witnessed the mining-armed conflict interaction. Communities where mining and armed conflict have converged can put the territorial peace approach to the test and illustrate the practical configurations of community action and mining dynamics in post-conflict environments.

Notes

1 The Colombian peasants who are increasingly understood as a cultural group rather than an ethnic group and whose rights are yet to be recognised in the Colombian context. At the time of the study, campesinos were likewise not recognised as a Census category in Colombia. For that reason, characterising their population proved difficult (see Duarte, 2015).
2 I use the term 'cosmogony' in part because it is a direct translation of the Spanish word used by the Nasa, but also because it emphasises the sacred source or origin of the Nasa world. The moral obligation to the mother-creator of the world is sacrosanct and understood as both a personal and communal relationship.
3 Owners of large extensions of land.
4 The study by Oxfam (Guereña, 2016) found that 0.4 per cent of farms were over 500 hectares and accounted for more than 67 per cent of productive land.
5 The abduction or imprisonment without authorisation of the State, accompanied with refusal to acknowledge the victim's whereabouts so as to remove him/her from the protection of the law (Henckaerts & Doswald-Beck, 2005). Forced disappearance does not always result in murder.
6 This is the 19th April Revolutionary Movement.
7 Popular Liberation Army.
8 Equitas stated the number of forced disappearances could double what the Centro Nacional de Memoria Historica has on record (Verdad Abierta, 2018).
9 Expert interview, 2017.
10 The Controller or *Contraloría* is the highest fiscal control body of the Colombian State. Its mission is to ensure good use of public resources and to contribute to modernising the State.
11 Interview with a Colombian mining and human rights specialist, 2017.
12 Although the purpose of expanding was set from earlier on, during the Uribe years (Ministry of Mines and Energy, 2006).
13 Since the enactment of Colombia's 2001 Mining Code all mining titles are granted as concessions, but prior to 2001 several types of titles existed. The word title therefore serves to refer to all mining titles regardless of when they were granted.
14 Including Anglo American Exploration Colombia, Anglo Gold Ashanti Colombia S. A., BHP Billiton's subsidiary Cerro Matoso, several Colombian, one Peruvian and one Brazilian company.
15 A review of the Montaño (2014) document prepared for ACIN indicates that despite selling its remaining interests in Anglo Gold Ashanti in 2009, Anglo American was still considered an affiliated company by communities, either through misinformation or because the divestment was not considered old enough.

References

ACIN. (2011). *Plan territorial cultural*. Santander de Quilichao, Colombia: ACIN.

ACIN. (2016a). *Historia de Nuestro Proceso*. Retrieved January, 2016, from www.nasaacin.org/sobre-nosotros2013/historia-de-nuestro-proceso.

ACIN. (2016b). *¿Quién lo conforma?*. Retrieved October, 2018, from www.nasaacin.org/sobre-nosotros2013/85-historia-de-acin.

ACIN. (2016c). *Sobre Nosotros*. Retrieved January, 2016, from www.nasaacin.org/sobre-nosotros2013/85-historia-de-acin.

Arbeláez-Ruiz, D., & Viana, A. (2015). *Análisis del Plan Nacional de Ordenamiento Minero de Colombia: Una lectura crítica desde la inclusión Social*. Brisbane: Centre for Social Responsibility in Mining, Sustainable Minerals Institute, The University of Queensland.

Bebbington, A., & Bury, J. (2013). Political ecologies of the subsoil. In A. Bebbington & J. Bury (Eds.), *Subterranean struggles: New dynamics of mining, oil, and gas in Latin America* (pp. 1–26). Austin: University of Texas Press.

Beltrán Rueda, L. E. (2016, April 27). *Nueve ríos de Cauca están en riesgo por la minería criminal El Tiempo.* Retrieved from www.eltiempo.com/archivo/documento/CMS-16575466.

Blundell, A. G., & Harwell, E. E. (2016). How do peace agreements treat natural resources? *Forest Trends Report Series: Forest Trade and Finance.* Retrieved from www.forest-trends.org/wp-content/uploads/imported/peace-agreement_natural-resources_formatted_final_1-19-16-pdf.pdf.

Bourdieu, P. (1977). *Outline of a theory of practice.* Cambridge: Cambridge University Press.

Cárdenas, H. S. (2018, July 5). *El mapa de los 311 líderes asesinados en Colombia El Colombiano.* Retrieved from www.elcolombiano.com/colombia/mapa-de-lideres-y-defensores-asesinados-en-colombia-DI8956261.

Centro Nacional de Memoria Histórica. (2016). *La maldita tierra: Guerrilla, paramilitares, mineras y con icto armado en el departamento de Cesar.* Bogotá, Colombia: Centro de Memoria Histórica.

Centro Nacional de Memoria Histórica. (2018). *Cifras: Los registros estadísticos del conflicto armado colombianoo.* Bogotá, Colombia: Centro de Memoria Histórica.

CINEP. (2012). *Minería, conflictos sociales y violación de Derechos Humanos en Colombia.* Bogotá, Colombia: Centro de Investigación y Educación Popular / Programa por la Paz.

Colombian Ombudsman. (2010). *Minería de Hecho en Colombia: Defensoría delegada para los Derechos Colectivos y del Ambiente.* Bogotá, Colombia: Colombian Ombudsman. Retrieved from www2.congreso.gob.pe/sicr/cendocbib/con4_uibd.nsf/F11B784C597AC0F005257A310058CA31/%24FILE/La-miner%C3%ADa-de-hecho-en-Colombia.pdf

Colombian Ombudsman. (2019). El riesgo de los defensores de derechos humanos merece mayor atención del Estado. *Defensor.* Retrieved January, 2019, from www.defensoria.gov.co/es/nube/noticias/7716/%E2%80%9CEl-riesgo-de-los-defensores-de-derechos-humanos-merece-mayor-atenci%C3%B3n-del-Estado%E2%80%9D-Defensor-Defensor-del-Pueblo-Carlos-Negret-Defensor%C3%ADa-derechos-humanos.htm.

Comisión de la Verdad. (2022). *Hay Futuro si Hay Verdad.* Volume 9. Resistir no es Aguantar. Bogotá, Colombia.

Consejo de Estado. (2016). *Asunto: Nulidad Electoral – Auto que admite la demanda y resuelve sobre la solicitud de medidas cautelares.* Bogotá, Colombia: Consejo de Estado. Retrieved from https://consejodeestado.gov.co/documentos/sentencias/19-12-2016_11001032800020160008100.pdf

Constitutional Court of Colombia. (2009a). Decision 004.

Constitutional Court of Colombia. (2009b). Decision 005.

Contraloría General de la República. (2014). *Análisis y evaluación. Plan nacional de desarrollo 2014–2018 "Todos por un nuevo país".* Bogotá, Colombia. Retrieved from www.google.com/url?sa=t&rct=j&q=&esrc=s&source=web&cd=13&ved=2ahUKEwiK-LXU6Z3eAhUHMI8KHcatBv4QFjAMegQICBAC&url=https%3A%2F%2Fwww.contraloria.gov.co%2Fdocuments%2F463406%2F472376%2FComentarios%2Bal%2BPlan%2BNacional%2Bde%2BDesarrollo%2B2014-2018.pdf%2Fe1781355-35fb-45da-9eea-0a12269d7778%3Fversion%3D1.0&usg=AOvVaw3hud_MWUjSvHg1zV3e-x44.

Delegates of the Government of Colombia and FARC. (2012). *General Agreement – General Agreement to end the conflict and build a stable and durable peace.* Havana, Cuba.

Delegates of the Government of Colombia and FARC. (2016). *Final Agreement to end the conflict and build a stable and durable peace.* Bogotá, Colombia.

Dougherty, M. (2011). *Peasants, firms, and activists in the struggle over gold mining in Guatemala: Shifting landscapes of extraction and resistance* (Doctoral dissertation, University of Wisconsin-Madison). ProQuest Dissertations Publishing.

Duarte, C. (2015). *Desencuentros territoriales: La emergencia de los conflictos interétnicos e interculturales en el departamento del Cauca.* Bogotá, Colombia: Instituto Colombiano de Antropología e Historia-ICANH.

Ebus, B. (2014, November 27). *Marcha de mujeres afro del Norte del Cauca llegó a Bogotá.* Retrieved February, 2016, from www.las2orillas.co/marcha-de-afrocolombianas-del-norte-del-cauca-llego-bogota/.

Echavarría, C. (2014). *What is legal? Formalising artisanal and small-scale mining in Colombia.* London: IIED and Colombia: ARM.

Fajardo, D. (2015). Estudio de los orígenes del conficto armado, razones para su persistencia y sus efectos más profundos en la sociedad colombiana. In Comisión Histórica del Conflicto y sus Víctimas (Ed.), *Contribución al Entendimiento del Conflicto Armado en Colombia.* La Habana: Comisión Histórica del Conflicto y sus Víctimas. Retrieved from https://indepaz.org.co/wp-content/uploads/2015/02/Version-final-informes-CHCV.pdf

Finji, M. T., & Rojas, J. M. (1985). *Territorio, Economia y Sociedad Paez.* Cali: Universidad del Valle.

Fundación Ideas para la Paz. (2013). *Dinámicas del conflicto armado en el sur del Valle y norte del Cauca y su impacto humanitario.* Bogotá, Colombia: Fundación Ideas para la Paz.

Gamarra Vergara, J. (2007). La economía del departamento del Cauca: Concentración de tierras y pobreza. *Documentos de Trabajo sobre Economía Regional*, 95.

Garay, L. J. (Ed.). (2013). *Minería en Colombia: Derechos, políticas públicas y gobernanza.* Bogotá, Colombia: Contraloría General de la Nación, Colombia. Retrieved from https://redjusticiaambientalcolombia.files.wordpress.com/2013/12/libro_mineria_contraloria-2013.pdf

Garay, L. J. (Ed.). (2014). *Minería en Colombia: Control público, memoria y justicia socio-ecológical, movimientos sociales y posconflicto.* Bogotá, Colombia: Contraloría General de la Nación, Colombia.

Giraldo, J. (2015). Aportes sobre el conflicto armado en Colombia, su persistencia y sus impactos. In Comisión Histórica del Conflicto y sus Víctimas (Ed.), *Contribución al Entendimiento del Conflicto Armado en Colombia.* La Habana: Comisión Histórica del Conflicto y sus Víctimas. Retrieved from https://indepaz.org.co/wp-content/uploads/2015/02/Version-final-informes-CHCV.pdf

González Rosas, A. M. (2016). *Vivimos Porque Peleamos – Una mirada desde abajo a la resistencia indígena del Cauca, Colombia.* Ciudad de México: Memorias Subalaternas.

Government of Colombia. (2010). *Plan Nacional de Desarrollo 2011–2014.* Bogotá, Colombia: Government of Colombia.

Government of Colombia. (2013). *Plan Nacional de Desarrollo 2014–2018.* Bogotá, Colombia: Government of Colombia.

Guereña, A. (2016). *Unearthed: Land, power and inequality in Latin America*. Oxfam International. Retrieved from www-cdn.oxfam.org/s3fs-public/file_attachments/bp-land-power-inequality-latin-america-301116-en.pdf
Guzmán Barney, Á., & Rodríguez Pizarro, A. N. (2015). *Orden Social y Conflicto Armado en el Norte del Cauca: 1990–2010*. Cali: Universidad del Valle.
Henckaerts, J. M., & Doswald-Beck, L. (2005). *Customary international humanitarian law* (Vol. 1). Cambridge University Press.
International Labour Organisation Convention on Indigenous and Tribal Peoples, Convention No 169 (1989) (ILO 169).
Jackson, J. E., & Warren, K. B. (2005). Indigenous movements in Latin America, 1992–2004: Controversies, ironies, new directions. *The Annual Review of Anthropology*, 34(1), 549–573. doi:10.1146/annurev.anthro.34.081804.120529.
Jaramillo, S. (n.d.). *Territorial peace. Office of the High Commissioner for Peace, Office of the President of Colombia*. Speech at Harvard University, Massachusetts.
Kearney, M. G., & Varese, S. (2008). Indigenous peoples: Changing identities and forms of resistance. In R. L. Harris & J. Nef (Eds.), *Capital, power, and inequality in Latin America and the Caribbean* (pp. 196–224). Lanham, MD: Rowman & Littlefield Publishing Group.
LaRosa, M. J., & Mejía, G. R. (2012). *Colombia a concise contemporary history*. Lanham, MD: Rowman & Littlefield Publishers.
Machado, M. (2015). *Ponencia en el Conversatorio Inclusión Social en los Planes y Políticas Públicos para el Sistema Minero de Colombia*. Paper presented at the Memorias del Conversatorio Inclusión Social en los Planes y Políticas Públicos para el Sistema Minero de Colombia, Bogotá, Colombia.
Massé, F., & Camargo, J. (2012). *Actores Armados Ilegales y Sector Extractivo en Colombia*. Informe V: CIT Pax, Colombia y Observatorio Internacional – DDR Ley de Justicia y Paz. Retrieved from www.catedras-bogota.unal.edu.co/catedras/gaitan/2016-I/gaitan_2016_I/docs/lecturas/s12/fmasse.pdf
Massé, F., & Camargo, J. (2013). Industrias extractivas y conflicto armado en Colombia. In A. C. González Espinosa (Ed.), *Los retos de la gobernanza minero-energética* (pp. 149–194) (Vol. 43). Bogotá, Colombia: U. Externado de Colombia.
Ministry of Education of Colombia. (2012). *Agropecuario Cauca: MinEducación*. Ministry of Education of Colombia. Retrieved from www.mineducacion.gov.co/1621/w3-article-299234.html
Ministry of Mines and Energy. (2006). *Plan Nacional para el Desarrollo Minero – Colombia País Minero*. Bogotá, Colombia: Mining and Energy Planning Unit, Ministry of Mines and Energy, Government of Colombia.
Ministry of Mines and Energy. (2016). *Política minera de Colombia- lineamientos para la minería del futuro*. Bogotá, Colombia: Ministry of Mines and Energy, Government of Colombia.
Molano, A. (2015a). Fragmentos de la Historia del Conflicto Armado (1920–2010). In Comisión Histórica del Conflicto y sus Víctimas (Ed.), *Contribución al Entendimiento del Conflicto Armado en Colombia*. La Habana: Comisión Histórica del Conflicto y sus Víctimas. Retrieved from https://indepaz.org.co/wp-content/uploads/2015/02/Version-final-informes-CHCV.pdf
Molano, A. (2015b). *A lomo de mula*. Bogotá, Cundinamarca, Colombia: Comunican SA.
Moncayo Cruz, V. M. (2015). Hacia la verdad del conflicto: Insurgencia guerrillera y orden social vigente. In C. H. d. C. y. s. Víctimas (Ed.), *Contribución al entendimiento del conflicto armado en Colombia*. Bogotá, Colombia: Ediciones desde abajo.

Montaño, M. (2014). *La Minería en Territorios Indígenas – base para la elaboración de lineamientos de política minera en territorios indígenas.* Santander de Quilichao, Colombia: Asociación de Cabildos Indígenas del Norte del Cauca.

Moor, M., & Sandt, J. v. d. (2014). *El Lado Oscuro del Carbón: Violencia paramilitar en la zona minera del Cesar.* Colombia. Utrecht, Netherlands: PAX Paises Bajos.

Morelli, S. (2014). Prólogo. In L. J. Garay (Ed.), *Minería en Colombia: control público, memoria y justicia socio-ecológical, movimientos sociales y posconflicto.* Bogotá, Colombia: Contraloría General de la Nación, Colombia. Retrieved from https://redjusticiaambientalcolombia.files.wordpress.com/2014/08/libro-mineria_contraloria_vol-iv.pdf

Mosquera, M. F. (2015a). *Ponencia en el Conversatorio Inclusión Social en los Planes y Políticas Públicos para el Sistema Minero de Colombia.* Paper presented at the Memorias del Conversatorio Inclusión Social en los Planes y Políticas Públicos para el Sistema Minero de Colombia, Bogotá, Colombia.

Mosquera, M. F. (2015b). *Ponencia en el Jornada de Trabajo Minería, Derechos y Territorio.* Paper presented at the Jornada de Trabajo Minería, Derechos y Territorio Santander de Quilichao, Colombia.

National Department of Statistics. (2005). *Censo Nacional.* Bogotá, Colombia: DANE.

National Department of Statistics. (2007a). *Colombia Multicultural: Su diversidad étnica.* Bogotá, Colombia: National Department of Statistics.

National Department of Statistics. (2007b). *La visibilización estadística de los grupos étnicos colombianos.* Bogotá, Colombia: National Department of Statistics.

National Department of Statistics. (2013). *Comunicado de Prensa – Cauca: Pobreza monetaria 2013.* Bogotá, Colombia: DANE.

Negrete, R. (2013). Derechos, minería y conflictos. Aspectos normativos. In L. J. Garay (Ed.), *Minería en Colombia: derechos, políticas públicas y gobernanza.* Bogotá, Colombia: Contraloría General de la Nación, Colombia. Retrieved from https://redjusticiaambientalcolombia.files.wordpress.com/2013/12/libro_mineria_contraloria-2013.pdf

Observatory of Human Rights and International Humanitarian Law of the Vice-president's Office. (2009). *Diagnóstico de la situación del pueblo indígena Nasa o Páez.* Bogotá, Colombia: Vice-president's Office, Government of Colombia.

Office of the High Commissioner for Peace. (n.d.). *Entérese del Proceso de Paz.* Office of the High Commissioner for Peace, Office of the President of Colombia.

Organization for Economic Co-operation and Development. (2015, January). Colombia: Policy priorities for inclusive development. *Better Policy Series.* Retrieved April, 2023, from www.oecd.org/about/publishing/colombia-policy-priorities-for-inclusive-development.pdf.

Ospina, W. (2013). *Pa que se acabe la vaina.* Spain: Grupo Planeta.

Otero Bahamon, S. (2006). *Los conflictos de autoridad entre los indígenas y el Estado. Algunos apuntes sobre el Norte del Cauca. La integración política y económica de los indígenas a la vida de la Nación: logros, retos y resistencias.* Retrieved February, 2016, from www.institut-gouvernance.org/en/analyse/fiche-analyse-340.html.

PBI Colombia. (2011). Minería en Colombia ¿A qué precio? *Boletín informativo,* 18.

Pérez Rincón, M. (2014). Conflictos ambientales en Colombia: Inventario, caracterización y análisis. In L. J. Garay (Ed.), *Minería en Colombia: Control público, memoria y justicia socio-ecológical, movimientos sociales y posconflicto.* Bogotá, Colombia: Contraloría General de la Nación, Colombia. Retrieved from https://redjusticiaambientalcolombia.files.wordpress.com/2014/08/libro-mineria_contraloria_vol-iv.pdf

Pizarro Leongómez, E. (2015). Una lectura múltiple y pluralista de la historia. In Comisión Histórica del Conflicto y sus Víctimas (Ed.), *Contribución al Entendimiento del Conflicto Armado en Colombia*. La Habana: Comisión Histórica del Conflicto y sus Víctimas. Retrieved from https://indepaz.org.co/wp-content/uploads/2015/02/Version-final-informes-CHCV.pdf

Political Constitution of Colombia. (1991).

Proyecto Nasa. (n.d.). *Marco Histórico del Pueblo Nasa*. Retrieved October, 2018, from www.proyectonasa.org/index.php?option=com_content&view=article&id=38&Itemid=215.

Rappaport, J. (2000). *La Politica de la Memoria*. Popayan, Colombia: Editorial Universidad del Cauca.

Redacción El País. (2015, August 21). PIB y exportaciones del Cauca crecen por encima de los actuales registros del país, *El País*.

Redacción Política. (2017). La cuestionable estrategia de campaña del No, *El Espectador*. Retrieved from www.elespectador.com/noticias/politica/cuestionable-estrategia-de-campana-del-no-articulo-658862.

Registraduría Nacional del Estado Civil. (2016). *Preconteo de votos del plebiscito*. Retrieved October, 2018, from https://elecciones.registraduria.gov.co/pre_plebis_2016/99PL/DPLZZZZZZZZZZZZZZZZZ_L1.htm.

Rettberg, A., & Ortiz-Riomalo, J. F. (2016). Golden opportunity, or a new twist on the resource-conflict relationship: Links between the drug trade and illegal gold mining in Colombia. *World Development*, 84, 82–96. doi:10.1016/j.worlddev.2016.03.020.

Supreme Tribunal of Antioquia. (2014). Specialis Civil Court for Land Restitution, Decision 2014–00005.

Supreme Tribunal of Bogotá. (2014). Justice and Peace Court, Decision 2006–80008.

Taussig, M. T. (2003). *Law in a lawless land: Diary of a "limpieza" in Colombia*. Chicago, Ill.: University of Chicago Press.

The Universal Declaration of Human Rights. (1948). United Nations. 217A.

United Nations Declaration on the Rights of Indigenous Peoples (UNDRIP). (2007). 61/295 C.F.R.

United Nations Office on Drugs and Crime & Government of Colombia. (2016). *Colombia: Monitoreo de territorios afectados por cultivos ilícitos 2015*. Retrieved April, 2023, from www.unodc.org/documents/colombia/2016/Julio/Censo_Cultivos_Coca_2015_SIMCI.pdf.

Urrea, F., & Viáfara, C. (2007). *Pobreza y grupos étnicos en Colombia: Análisis de sus factores determinantes y lineamientos de políticas para su reducción: Misión para el diseño de una estrategia para la reducción de la pobreza y la desigualdad (MERPD)*. Departamento Nacional de Planeación.

Valencia, L., & Riaño, A. (2017). *La Minería en el Posconflicto: un asunto de quilates*. Bogotá, Colombia: Editorial B.

Van Cott, D. L. (2001). Explaining ethnic autonomy regimes in Latin America. *Studies in Comparative International Development*, 34(4), 30–58.

Vega, R. (2015). La dimensión internacional del conflicto social y armado en Colombia. In Comisión Histórica del Conflicto y sus Víctimas (Ed.), *Contribución al Entendimiento del Conflicto Armado en Colombia*. La Habana: Comisión Histórica del Conflicto y sus Víctimas. Retrieved from https://indepaz.org.co/wp-content/uploads/2015/02/Version-final-informes-CHCV.pdf

Velasco, M. (2011). Contested territoriality: Ethnic challenges to Colombia's territorial regimes. *Bulletin of Latin American Research*, 30(2), 213–228. doi:10.1111/j.1470-9856.2010.00500.x.

Verdad Abierta. (2014). Los pueblos étnicos del Cauca piensan en el posconflicto. *Verdad Abierta*. Retrieved from https://verdadabierta.com/los-pueblos-etnicos-del-cauca-piensan-el-posconflicto/.

Verdad Abierta. (2015a). ¿Habrá zonas de reserva campesina en Valle y Cauca? *Verdad Abierta*. Retrieved from https://verdadabierta.com/habra-zonas-de-reserva-campesina-en-valle-y-cauca/.

Verdad Abierta. (2015b). *La petroguerra del ELN en Arauca Verdad Abierta*. Retrieved from https://verdadabierta.com/la-petro-guerra-del-eln-en-arauca/.

Verdad Abierta. (2018). Desaparecidos en Colombia podrían ser el doble de los registrados por el CNMH. *Verdad Abierta*. Retrieved from https://verdadabierta.com/desaparecidos-en-colombia-podrian-ser-el-doble-de-los-registrados-por-el-cnmh/

Viana, A. (2015). *Mapa de Un Sistema Minero Incompleto. Informe de análisis de contexto preparado para el proyecto Promoviendo la Inclusión Social en la Planeación Minera Nacional*. Brisbane, Australia: Centre for Social Responsibility in Mining, The University of Queensland. Retrieved from www.csrm.uq.edu.au/publications/mapa-de-un-sistema-minero-incompleto

Weitzner, V. (2012). *Rendición de cuentas de las compañías extractivas en Colombia: Una evaluación de los instrumentos de responsabilidad social empresarial (RSE) a la luz de los derechos de los Indígenas y los Afrodescendientes*, trans. M. C. Benítez (p. 179). Bogotá, Colombia: The North-South Institute, Proceso de Comunidades Negras, Resguardo Indígena Cañamomo Lomaprieta.

World Conference Against Racism, Racial Discrimination, Xenophobia and Related Intolerance Declaration. (2001). UN Doc. In *A/CONF* (Vol. 189, No. 12, p. 8).

3 An Immoral Economy

Introduction

As we prepared for the work I was going to do in Canoas, we held a workshop with the local Nasa researchers who knew the mining dynamics in detail. We met at a Harmonisation Centre, where Indigenous people who have 'brought disharmony' by disobeying Indigenous law spend time being re-harmonised. Usually training events are held there to make the most of those opportunities. We started with me talking about the main aspects of the research and what I was hoping to achieve so that the others could frame their guidance, questions and observations. I breezed through the nuts and bolts, as these were experienced researchers. Then I brought up the idea of the 'moral economy'. The Nasa researchers swiftly replied along the lines of "*Do you mean the IMMORAL economy of mining?*" and that led to detailed explanations of the negative consequences of mining. So, even in discussing concepts, it was important that there be no ambiguity – mining was immoral to them.

The Nasa have an intellectual tradition that has been vital in elucidating their history, cosmogony and organisational process (Rappaport, 2000). Rappaport documented the workings of Nasa intellectuals and historians dating back to colonial times and identified how their discourse provided the concepts and grounding for contemporary Nasa intellectuals to critically analyse Nasa ideas (Rappaport, 2000). This tradition has also integrated collaboration between Nasa and academic intellectuals (Rappaport, 2020). For example, an intercultural research, reflection, education and learning space for Nasa women called *Tulpa del Pensamiento de las Mujeres Nasa y el Territorio* provided a space for women to look at their roles and their future aspirations and then identify and implement actions as a collective, allowing them to develop an understanding from their standpoint as Nasa women (Llano Quintero, 2012). Through a research process that builds on existing Nasa resources and traditions, the Nasa have developed a standpoint that permeates through their concepts and principles, their cosmogony and everything their organisations do.

The Nasa researchers I was working with this time in Canoas were quick to set the record straight at the beginning of our collaboration, because mining is

DOI: 10.4324/9781003226895-3

the subject of strong political, economic and moral tensions between Indigenous organisations and the supporters of mining activities. The latter include the mining industry, the supporting State apparatus of entities in charge of promoting and protecting formal mining investment and development, informal or illegal miners, and the miners who associate themselves with illegal armed groups. There is significant tension amongst the Nasa over the threat of illegal mining and of mining with links to criminal groups. Where formal mining is concerned, a central source of tension is that the 1991 Constitution gives ownership of the subsoil to the State. In parallel, it grants territorial autonomies to Indigenous communities and requires that they take part in decisions regarding the use of subsoil resources. Territorial autonomy presents one of the most difficult interactions between the National Government and Nasa Authorities. The tensions and concerns amongst the Nasa relate as much to formal mining, and therefore to interactions with State and industry, as they do to informal mining, illegal or mining activity with links to criminal groups. At the time of the study there were few formalised mining ventures in North Cauca, and in Cauca, overall, one of the largest concerns was the proliferation of illegal mining activity and mining with links to criminal groups.

Earlier, I contended that when it comes to mining by outsiders in Nasa territories, a prevalent position among Nasa participants was a NO to mining. Here, I discuss the explanations, narratives and feelings about mining held by Nasa participants who oppose mining. These participants appealed to moral principles, to their expression in Indigenous mandates and law, to moral tensions and dilemmas, to what a good Nasa ought to do, and to how mining affects the Nasa world. Based on this emphasis on morality, the analysis I present here is a sketch of the contours of the Nasa of North Cauca's moral economy as it relates to mining. The moral economy refers to the customary norms about how markets should operate that are specific to a time and place (Owen, 2009). Here, I describe how outsider mining was seen within the Nasa moral economy in the Colombian Peace Agreement transition of 2016 and 2017.

A protagonist in this description of mining within the Nasa moral economy is the natural world that the Nasa cosmogony describes as made up of animate beings, spirits and people linked through strong reciprocity and affective relationships that resemble organic links. I use the term 'cosmogony' in part because it is a direct translation of the Spanish word used by the Nasa, but also because it emphasises the sacred source or origin of the Nasa world. The moral obligation to the mother-creator of the world is sacrosanct and understood as both a personal and communal relationship. The elements of the natural world such as trees, water, fish, mountains or gold itself are, in the Nasa cosmogony, actors with capacity to affect people not only physically but morally and affectively. All of them contribute to shaping this moral economy. The Nasa moral economic thinking delineates an organic perspective of the economy where correct actions equate to life-sustaining reciprocity and transgressions sever vital reciprocity.

This instilling of nature with morality and, mediated through reciprocity obligations, the consequent role of indigeneity as "*moral anchor*" (Horst, 2019) reflect what has been documented elsewhere in Latin America (Ødegaard & Rivera Andía, 2019). In extraction contexts, Indigenous responses become a site to explore Indigenous ontologies and cosmogonies (Ødegaard & Rivera Andía, 2019), as much as a space to make indigeneity visible in opposition to extraction (Li, 2015). While my exploration concerns the Nasa during the post-conflict transition in Colombia, others have studied Indigenous resistance in the context of extraction at multiple sites to offer a regional perspective (see Ødegaard & Rivera Andía, 2019) or examined Latin American Indigenous identity and resistance dynamics despite considering it "*a daunting task*" in this diverse region (Kearney and Varese, 2008).

I shall discuss the Nasa participants' prevalent view that most forms of mining, and in particular those that outsiders practice in Nasa territories, are not aligned with a good life or what being a good Indigenous person entails. The concept of Buen Vivir (Living Well) or, in NasaYuwe, *Wët wët fizenxi*, embeds the dreams of the Nasa communities about a good life and permeates moral discourse. Guidance on how to live to attain those dreams comes with the movement's moral principles of Unity, Land, Culture and Autonomy – *Unidad, Tierra, Cultura y Autonomía*. The movement's mandates and discourses are explicitly built around these principles and embed the ideal of *Wët wët fizenxi*. In light of these principles and ideas, many in the North Cauca Indigenous movement see various forms of mining as moral failures. The prevalent discourse speaks of a clash between mining and the movement's moral framework, a clash that the participants described through what I shall refer to as 'narratives of mining-induced degradation'. Through these narratives I learnt about the spiritual, social, economic, environmental and political degradation that these participants associate with mining.

Beyond the thick contours that the idea of Buen Vivir and the principles of Unity, Land, Culture and Autonomy provide, there are other moral 'lines' that appear more tenuous, but no less relevant. By virtue of the material conditions and historical and systemic discrimination and disadvantage of Indigenous people, some values might not warrant explicit statements to guide the work of Indigenous organisations because their validity 'comes with the territory'. Ideas of fairness and resistance permeate the Nasa moral discourse on mining that I discuss in this chapter. While I shall concentrate on the thick lines, I shall also reflect on these ideas along the way, to provide a closer approximation to the moral landscape within which positions towards mining form.

Buen Vivir, Living in Harmony

The idea of Buen Vivir guides Indigenous communities to promote harmony with the self, within the family, the community and nature. It is a prominent concept in public materials from Cauca Indigenous organisations such as the Autonomous Intercultural Indigenous University or UAIIN for its Spanish acronym (see Universidad Autonoma Intercultural Indígena (UAIN-CRIC), n.d.). Buen Vivir speaks of balanced and harmonious relationships in those domains. Within this vision or ideal of life, the relationship with nature is embedded rather than binary. Buen Vivir refers to collective wellbeing – better living conditions for all rather than the individualistic pursuit of a higher standard of living. In Luis's (pseudonym) words:

> *Wët wët fizenxi says that Buen Vivir is for you to have decent housing, for you to have agricultural production, for you to have a good relationship with the community and above all with nature. Buen Vivir is like living in peace with the community, with each other, with the family, with your own nature, and for you to have sustenance. That means Buen Vivir is different from living better (vivir mejor). Living better is living independent of the rest of the community, even if they starve, while Buen Vivir is Buen Vivir for the entire community.*

Defenders of Buen Vivir refuse to refer to the place they inhabit in terms of resources to exploit in a transactional manner, or to adopt ideas of unlimited, linear economic growth that promote individual accumulation at the expense of community wellbeing. Rather than accumulation and growth, Buen Vivir advocates propose reciprocity, fulfilment, self-realisation and achieving freedoms (Acosta, 2008; Consejo de Mujeres Originarias por el Buen Vivir, 2015; First Congress of the Black Peoples of Colombia (CNOA), 2014; Gudynas, 2011). This dichotomy between Buen Vivir and living better is based on the understanding that there are trade-offs between: capitalistic accumulation, with mining as an integral part of that system; and collective wellbeing and harmony.

Some translate *Wët wët fizenxi* as living beautifully (*Vivir Bonito*) in a harmony that contrasts the selfishness the Nasa observe in capitalism. This encompasses living freely or walking freely, in harmony with Mother Earth and all the beings on it. *Wët wët fizenxi* requires consulting with the spirits, asking for permission and making offerings before any disruption of the land or the landscape. These actions seek to maintain harmony. Looking after the territory, growing crops, caring for children and contributing to the community increase a Nasa's moral worth. On the contrary, anything that degrades the land or the territory, hurts Mother Earth or harms the community and family is disapproved of. Kearney and Varese (2008, p. 201) observed a "*moral management of the cosmos, a type of environmental ethic and practice found in the majority of the 'indigenous societies'*" of the Latin American region and a moral economy informed by this "*ecological cosmovision*" that has "*guided indigenous resistance to exploitation and political oppression*" (p. 204). Most Nasa I interviewed see a combination of capitalism with extraction and natural devastation as an economic model of selfishness that the Nasa ought to resist, because it limits people's freedoms, through exploitation of both people and nature and through the dearth of rural livelihood options that comes with environmental degradation.

These ideas clearly situate the Nasa in the development and alternatives-to-development debates in Colombia and the Latin American region. In Colombia, extractive industry growth has been prevalent in National Government development policy, while many regions and communities have strong agricultural identities and attachments to those cultural landscapes and see extraction as a threat (Government of Colombia, 2010; Negrete, 2013; Rodríguez-Franco, 2014). Reflections of these attachments include the popularity of citizens' consultations to block mining projects and court decisions that support local and regional actors to participate in extractive related decisions (Constitutional Court of Colombia, 2016a; Rodríguez-Franco, 2014). As Eduardo Gudynas (2017) observed in Latin America, "*the debate about the merits and limits of extractivism occurs live in prime time television, it is not a debate between the greenies*".[1] It is very much a mainstream debate. Increasingly rivers and other ecosystems are acquiring legal rights, or protection as beings in some of the national constitutions. The fault lines in this debate on extraction do not mirror right and left divisions. Governments across the political spectrum have driven an expansion of the extractive frontier and the region has consolidated its role as a supplier of raw materials (Bebbington, 2009; Bebbington & Bury, 2013; Gudynas, 2009). The debate on extraction is about the role, if any, that it might have in the desired futures of people's and nations. Defenders of Buen Vivir advocate it as an alternative path to the problems of capitalism and extraction (Acosta, 2017; Warnaars & Bebbington, 2014).

The moral positions behind Buen Vivir are connected to global debates (Gudynas, 2011). For example, it is well acknowledged that the discourses of Buen Vivir and Deep Ecology have cross-pollinated. Buen Vivir is consistent with the concept of 'self-realisation' that Norwegian environmental thinker and proponent of Deep Ecology, Arne Naess, defends. In Naess's self-realisation, the self "*embraces all the life forms on the planet [...] together with their individual selves*" (2005, p. 271). Naess states that self-realisation implies "*maximising the manifestations of all life*" and exercising a principle of "*live and let live!*" where "*the self is extended and deepened as a natural process of the realisation of its potentiality in others*" (p. 272). Under this understanding, wellbeing can only be pursued collectively and in harmony with nature. This resonates with the Nasa explanations of Buen Vivir and illustrates the global influence of these Indigenous ideas.

Moral Tensions and Mining-Induced Degradation

For many participants, practicing mining is not Buen Vivir because the advance of any form of mining in Indigenous territory goes against the possibility of Buen Vivir. The Nasa views on mining that I discuss in this chapter reflect concerns that there are many forms of mining that extinguish life and take away from nature, people, communities and families without reciprocity or harmony. The central view is that such mining activities compromise the

Indigenous communities that are one with the land, dividing their loyalties. Or they may damage places of spiritual significance and damage the water and land needed to grow food. There is concern that dependence on the cash economy sees Indigenous people purchase food from outside the community. For the participants, this is a loss not only for the local economy, government and autonomy, but also for community health as the diets become less balanced. The violence of illicit armed groups associated with illegal mining threatens the lives and freedoms of Indigenous people. The list of grievances and moral concerns is lengthy and leads people to see mining as an obstacle to Buen Vivir or *Wët wët fizenxi*. These obstacles and moral tensions are best understood by considering the Nasa moral principles. It was around enumeration of those principles that participants discussed the narratives of mining-induced degradation that I elaborate on in the following sections.

Overall, the moral principles of Unity, Land, Culture and Autonomy configure a moral order that favours the ideal of a rural way of life in an intricately connected and harmonious community. Not individuals but communities, acting in unison without division, with strong mutual obligations and networks of reciprocity. People should be committed to a role of stewardship over the land to which they are firmly and definitely attached. They should make autonomous, localised decisions that prioritise the collective good as well as stewardship obligations. They ought to reproduce the Nasa way of life, culture and ways of understanding the world. Implicit in these principles come the ideas of fairness, for example, in rewards for hard work and in access to land; and of resistance, a call that is never presented as optional for members of a historically oppressed community that has won recognition through a long and difficult process. Referring to Latin American Indigenous people in general, Kearney and Varese (2008, p. 202) describe such a moral economy thus:

> *It seeks to preserve the common resources of the community and minimise internal economic differentiation. Within such a community, basic economic resources are held in common, and access to them is determined by good citizenship, defined in terms of willingness to serve the community in ways that often involve considerable self-sacrifice. Such an economy is centred more on use value than on exchange value, and economic transactions are mediated primarily by reciprocity rather than profit considerations.*

It is clear, however, that the Nasa's material situation requires engagement with mainstream capitalist markets and that they ought to balance reciprocity with market considerations. Nevertheless, this description is a useful one that illustrates the ideals of their moral economy.

The economic, social, spiritual, political and environmental changes that occur with mining create moral tensions. First is the tension between individualism and collectivism. Miners, be they local Indigenous people, formal mining companies or illegal or criminal mining ventures, are seen to deplete the water and land that others' livelihoods depend on. In the eyes of many, once there are Nasa miners, opposing miners external to the territory, legal or illegal, becomes harder. There is tension between growing, making and sharing on the one hand, and buying on the other; between creative forces, and a form of surrender that many Nasa equate with a culture of easy profiteering. That is in itself a tension, one between long-term endeavours, with gradual and hard-earned rewards, and high-risk activities, with potentially significant monetary short-term gains. This reveals another tension – embracing an economy based on commodification, or remaining in an economic framework predicated on existing relationships and power networks. As consumption increases, cultural tension ensues between maintaining Indigenous culture or opting for what is seen as an imported popular culture. Some Nasa believe the younger generations might be too partial to the promises and pleasures of consumption. These tensions mirror the perceived duality between differentiation and mainstreaming that the Nasa face. But the tensions do not stop there. In a context of protracted armed conflict with pervasive illicit economic networks, the means to exert violence are all too available and tensions about mining can escalate. They have, at times, become matters of life and death.

All these tensions were evident in the participants' narratives of mining-induced degradation. The premise is that the Nasa way of life is in danger whenever people fail to observe the moral principles. The concern is that when Indigenous people get involved in mining, they are either unable or unwilling to uphold the movement's principles. Most Indigenous participants, not involved in mining, associate it with social, cultural, spiritual, economic, political, moral and environmental degradation in Indigenous communities. They see rapid changes within the communities as mining arrives, beginning with miners challenging Traditional Indigenous Authorities. Many attribute to mining the same social ailments that accompany illegal economies such as crops of illicit use: a toxic mix of gender impacts, proliferation of weapons, arrival of and recruitment by illegal armed groups, unexplained deaths, violent and sometimes lethal disputes, abundance of festivals and rising alcohol consumption, environmental degradation, loss of agricultural identity, loss of commitment towards the community and degradation of the Indigenous identity. Let us examine how mining and the moral principles interact in the eyes of the participants.

The participants' narratives present a set of interconnected principles where failure to abide by one extends failure to some or all other principles. However, for clarity, here I organise the narratives under each of the principles, analysing the stories that speak most strongly to the idea of degradation that people perceived as indicative of moral failure in relation to each principle.

Unity

The Unity principle calls for Indigenous people and communities to support one another and pursue causes as a coherent block, without divisions, building loyalty, and emotional and strategic strength and safety through numbers. Unity is an antidote to the fear that comes with living during intense armed conflict. It is a strategic advantage against rival external interests, and a source of protection. Unity makes resistance, and therefore survival, possible and, in doing so, supports the pursuit of Indigenous Life Plans containing territorial, social, spiritual, political and economic objectives. Resistance often takes the name of re-existence in Indigenous and Afro-descendant circles. It refers to:

> *The devices that communities create and develop to invent life in the day-to-day and be able to confront a hegemonic project that has silenced them, portrayed them as inferior, and made them invisible [...]*
>
> *The organisation, production, nutrition, ritual and aesthetic codes that can dignify life and re-invent it so that communities can continue to transform.*
>
> (Albán Achinte, 2009, p. 455)

These codes cannot be maintained or invoked successfully without a collective community belief and commitment to their power and currency.

Unity comes in narratives of strength in numbers. For example, the Indigenous Guard's anthem tells us how younger people, in the thousands, will come to take the place of those who perish in the struggle. Sometimes the appeal to Unity seeks to boost morale. It lends confidence that the collective effort will not be in vain. For example, Pedro (pseudonym) explained to me that "*if you are on your own, they can kill you, but if it's all of us, they might kill ten or twenty, but they can't kill us all, we will disarm them in the end*". Large numbers signify support and provide leaders the emotional endurance and the courage to assume great personal risk to stand up to groups harming the community. Pedro spoke about Unity being the source of strength to confront a gang of violent thieves saying: "*had I been on my own, I wouldn't have had the strength, but with 300 comuneros backing me, I had the courage to confront them*". Unity creates ties of loyalty that protect Indigenous people. I heard many times that community members refused to give away leaders' whereabouts to murder squads thus saving the leaders' lives. Unity becomes power in a political struggle that often resembles a game of attrition. It provides individual strength and offers each community member a shield.

The moral principle of Unity reflects the needs of an Indigenous resistance faced with violent repression to its collective struggle for rights, for land and for a political project. To survive, to be effective and to be visible, Nasa resistance relies on cohesion and numbers. From its beginnings, through 'land recovery' efforts, the Indigenous movement aligned itself with the rest of the rural poor, or campesinos, resisting long established, pervasive patterns of exploitation (González Rosas, 2016; Otero Bahamon, 2006; Velasco, 2011).[2] Indigenous resistance and land attachment have confronted the territorial control orders of illegal[3] armed actors, such as guerrillas, drug traffickers and criminal bands (González Rosas, 2016). In resisting the prevailing rural economic order and disparate violent projects of territorial control, the Nasa have faced violent, persistent repression by State and non-State actors (González Rosas, 2016; Guzmán Barney & Rodríguez Pizarro, 2015). Unity as a collective is crucial for this resistance to be effective. The list of murdered Indigenous leaders and community members in North Cauca is lengthy (see González Rosas, 2016). Only a movement with strong numbers can survive such relentless attacks against the lives of its members without being completely destroyed.

The idea of Unity encapsulates reciprocal obligations to fellow Indigenous people past, present and future. Behind Unity is the notion that Indigenous people will move[4] into the future through the resistance they are contributing to now. Likewise, contemporary Indigenous people recognise their current rights and territories as the result of the sacrifice of previous generations. Consequently, what is expected of them is a willingness to invest and sacrifice themselves in a similar fashion for future generations. Cxayuce (pseudonym), a Nasa leader, sees the relationship between generations like this: "*White people think the children and the future generations are ahead, but for us they are behind. Children walk behind us, we are walking ahead of them and preparing the path for them*". In the Unity principle, Nasa organisations set a standard of intergenerational duty that supports continuity for their collective efforts.

This Indigenous movement and its organisations consider self-sacrifice for the sake of the collective a moral duty of their members. This makes the pursuit of individual gain reproachable, in particular if it deviates from collective goals. Adversarial relationships between the movement and powerful sectors of the mainstream Colombian society see the movement targeted with 'divide and conquer strategies' and with a discourse that portrays indigeneity as an obstacle to development and economic growth, and Indigenous social mobilisation in Cauca as a process infiltrated by illegal armed groups (Calderón, 2018; see for example Piedrahita, 2016; Rico, 2009). The Nasa are heavily stigmatised in Colombia and many share a feeling of lack of domestic support (Rappaport, 2000). Because the Nasa mostly rely on each other, individualistic pursuits or activities that divide the communities are seen as eroding a fundamental basis for collective strength.

So crucial is the duty of self-sacrifice for Indigenous resistance that the discourse places no limits on the acceptable personal costs. The anthem of the Guardia encourages people to resist and defend their rights. It speaks of how younger generations will come to take the place of those who perish in the struggle. Willingness to stand up for the movement's causes, despite violent repression, even risking one's life, is a normalised element in debates and public speeches, commonplace in social mobilisations such as protests and in discussions about land rights. Statements to that effect do not generate any noticeable reaction. When talking with me about resisting anything from various forms of mining to drug trafficking, the armed conflict, violent Government repression or violence from criminal actors, many of the participants volunteered that they were willing to sacrifice their lives in the struggle. I was surprised that they would say this unprompted when I was not directly inquiring about the limits of their resistance or the personal risks. Such gestures, normalised and commonplace as they might appear, are neither hollow nor superficial. There was no heroism or bravado. On the contrary, I was able to appreciate why the Nasa are known for their stoicism. Instead, the statements reflected sadness and fatigue over the persistent violent threats and ongoing killings. Only once did I hear people question the idea of sacrificing one's life. In Corinto's rural area, regular confrontations between police squadrons and the Nasa occur and have led the locals to reflect on the risk of facing violent death when protesting.

Unauthorised Mining Arrives and Unity Crumbles

When a dispute over access to an unauthorised mine led one Indigenous miner to kill another in Munchique, Nasa communities and organisations were concerned that mining goes against Nasa Unity. Not only did the miners overlook the mandates of their Indigenous Authorities, they escalated the dispute into violence against fellow Indigenous people. Talking about yet another confrontation between Indigenous people over a different unauthorised mining site, Pedro described to me his feelings as he saw the tensions escalate close to violence. He said to me with profound sadness, "*That day I thought: This is not possible, I cannot end up being killed by another Indio, by one of my own*". The call to Unity is to sacrifice for each other, not to sacrifice one another as in these mining disputes.

Pedro's sadness is a shared feeling. Nasa participants displayed most sadness and apprehension when they spoke about Indigenous people allowing divisions about mining to grow and cause clashes with fellow community members. Sek (pseudonym) said to me: "*It is easy to confront a foreign mining company, but what is really difficult is to be put against our own fellow comuneros*". The Nasa are all too aware that, as Scott (1985, p. 302) explained, "*certain combinations of atomization, terror, repression, and pressing material needs can indeed achieve the ultimate dream of domination: to have the dominated exploit each other*" and resort to "*beggar-thy-neighbour strategies*". Where these conditions are largely present, what is within the Nasa's control is resisting atomisation, and what better proof that one does not "*beggar-thy-neighbour*" than self-sacrifice for the collective. With mining, sacrifice continues but loyalties shift. This is in itself a source of fear because it corrodes people's sense of belonging, of being included and of identity. As Tilly (2005) theorised, identities reside in relationships within and across boundaries of 'us and them', providing the basis for acts of inclusion and exclusion. Mining-driven divisions between the Nasa indicate severed relations and emerging forms of exclusion.

Mining Can Intensify Divisions

Indigenous people need the collective visibility that Unity can offer because their opponents are avid to exploit signs of division to campaign against

Indigenous rights. Indigenous people have legal rights and need – if not legally because legal recognition is already there – at least in the realm of public perception, evidence of ongoing presence and of a consolidated Indigenous position. So the Indigenous resistance can be performative (González Rosas, 2016), to some extent, as a response to sections of the mainstream society questioning Indigenous rights and calling for homogenisation on the basis of Colombia's mixed background majority.

According to several participants, external proponents of mining in Indigenous territories seek entry by co-opting community members and leaders, exploiting poverty to divide a community that would otherwise uniformly oppose mining. Some see mining as a way out of poverty, others see in it an empty promise of prosperity that sets off a vicious cycle of debt, exploitation and violence. Miners might be a minority, but they contest the discourse of the Indigenous movement.

Critics from conservative sectors of the media, industry and politics prey on divisions within Indigenous organisations to argue against legal rights, in particular autonomy. Curiously, one of the sites of these publicised divisions is a dissident Nasa community that, at the time of the study, was undertaking and defending mining. In 2016, a mainstream media outlet was found guilty and forced to rectify and apologise to Colombia's most prominent Indigenous organisations, including Cauca's, over attacks aimed at eroding public support for Indigenous autonomy (Constitutional Court of Colombia, 2016b). The Courts found that the series of primetime television shows, devoted solely to 'dysfunctional' Indigenous communities, were intentionally defamatory and stigmatising. The Nasa site profiled in the media was El Nilo, in Caloto. It was the site of the massacre (see Chapter 2, p. 31), the subject of collective compensation, the site where divisions and victims' grievances soared about the use of those moneys and where many Indigenous people had chosen to become miners, where they had decided that they did not belong to the well-established Cauca Indigenous organisations, but to a new organisation that failed to get recognition by any Colombian institution. It was at El Nilo where Julian (pseudonym) saw an image he dared me to imagine. Picture this, he said: "*an Indigenous man climbed onto a digger and shouted to the Guardia*: *you can burn me with the machine*!" This was effectively setting limits to the Guardia's law enforcement and to the rest of Nasa Authorities in North Cauca.

El Nilo was a site of political division and there were Indigenous miners there. But in this case, it is unlikely that mining alone precipitated the fissure in the Indigenous movement. However, at sites of political contestation, mining appears to provide a critical fuel. It yields the funds to support alternative leaders in their political and economic contentions. As Sek said, referring to Canoas, "*We need to understand whether mining is something that people use to sustain disputes with the Cabildo*". The Mining Analysis group of Canoas had also come to similar reflections (Caro Galvis & Valencia, 2012). Mining is a suitable choice for that purpose. Not only does it send a message of dissent and economic discontent, it also provides the funds to harness power and local support.

Land, Ancestral Territory or Mother Earth

Land, territory[5] or Mother Earth (*Madre Tierra*) constitutes a fundamental basis of Nasa life. Land is the source and sustenance of everything and every aspect of life. Several participants explained to me that Indigenous life is contingent on Mother Earth because Indigenous people are part of Mother Earth; they cannot be separated from her or harm her. Harming the land – for example, using explosives, polluting, killing or affecting other living and sentient beings indiscriminately – is morally corrupt and is punished. Losing land, or giving way to the economic and political forces that take land from Indigenous people, goes against the Indigenous movements' objectives. So, when someone does not support the 'Liberation of Mother Earth' campaign, other Nasa infer a lack of alignment with shared objectives and a morally incorrect position. The Land principle embeds a network of interconnected ideas and moral obligations that reflect Nasa cosmogony and knowledge, objectives and dreams that influence all aspects of life.

Three discernible elements within the Land principle influence Nasa moral understandings of mining: i) land as the economic basis of subsistence and freedom; ii) land as a territory that represents a space for existence past, present and future; and iii) land as a sentient, motherly being: Mother Earth. Here, I shall consider these three components I observed in the Land principle and illustrate the reasons for moral disapproval of mining among the participants.

Land as the Material Basis of the Nasa Economic System

First, land is the basis of the economic life and struggles of the Indigenous people. It is the basis of their freedom from servitude. It was from the land struggles of the 1960s and 1970s that the Cauca Indigenous movement, Colombia's first, emerged. Land was what the rural proletariat fought for, in many cases, guided by a Marxist understanding of the economy. In those fights by the campesinos, Indigenous leaders were prominent. Their separation into an Indigenous movement happened later[6] (González Rosas, 2016; Taussig, 2010). As Scott (1977) and Moore (1978) have illustrated in other contexts, moral-economic grievances can lead rural workers to participate in resistance movements to change a rural economic order that breaches their criteria of fairness and hinders their subsistence. Indigenous people were part of the land struggles because land is the basis of Indigenous survival in its most material, immediate sense of providing food security. Freedom from servitude was difficult to achieve without community land and without an end to colonial era tributes or *terraje*. [7] The latter did not begin materialising until the 1970s, following legal reform in the 1960s (Velasco, 2011) and with the land recuperation achievements of the Indigenous movement (García, 2017; Pachón, 2005). Consequently, Consejo Regional Indigena del Cauca's (CRIC, n.d.) platform reflects the priority of securing land including: i) "*recovering the land of Indigenous reserves*" ii) "*expanding Indigenous reserves*" and iii) "*not paying tributes to landowners*" (or terraje). The aims are to improve the material conditions, securing the basis for the economic system and enhancing freedoms.

The Nasa experience with mining, both historical and contemporary, has created strong negative associations related to dispossession, environmental degradation and exploitation of the Indigenous people. Pedro's experience was that, "*Once mined, the soil never goes back to what it was. Mining washes away the soil, leaving only the rock. Not even the shrubs for a broom can grow in that soil*". To him, people from outside the community come to mine without any regard for the consequences for Indigenous material sustenance. Néstor (pseudonym) told me how, in Corinto, there was mining during colonial times and the land was taken from the Indigenous people. "*So now we know there is gold but we are keeping it secret, we are not saying where it is*", he said. There was a time at the turn of the eighteenth century when Cauca was fundamentally a mining province, there was little agriculture and food came mainly from livestock farms (Finji & Rojas, 1985). Mining provided the resources for *criollo* (American of Spanish descent) families to continue to expand their territorial power. Indigenous people were pushed to the boundaries in the higher reaches of the mountains. Indigenous memory deeply embeds an understanding that mining has been a vehicle of dispossession and wealth concentration.

Territory

Land as Territory or Ancestral Territory is the space where all forms of life, spirits and places of spiritual significance exist. It provides opportunities to reproduce the Nasa world into the future, or to persist. It is the space where the Nasa have developed a sense of place and contains the elements of their day-to-day life. Nasaus (pseudonym) told me the territory "*is the big house where we all live*", while Martin (pseudonym) explained it this way:

> *So we have recovered the land, but we did not recover her just for ourselves, we must defend life, and life is in the plants and in the sacred sites, in the animals and so on. We are not her only inhabitants, there are other beings who inhabit her, so we speak about the land because we need it, but we talk about a much wider concept covering all who can live in the territory.*

This understanding of territory has parallels with the idea of intrinsic value, that Deep Ecology enunciates thus: "*The well-being and flourishing of human and non-human life on Earth have value in themselves. These values are independent of the usefulness of the non-human world for human purposes*" (Naess, 2005, p. 239). The concept of Ancestral Territory recognises those intrinsic values, including the value of connections between the beings that exist in the territory.

The Territory is a domain that enables identity and memory. Protecting the Ancestral Territory is part of CRIC's ten-point platform (CRIC, n.d.). Territory refers to a space that makes it possible to live and to remember as an Indigenous people. It has embedded, and been the means to recount, Nasa memory and history since the early Nasa historians (Rappaport, 2000). The Nasa favour a cyclical notion of time, which refers back to the landscape and specific sites. The territory provides a form of sustenance beyond the material, enabling the ritual and spiritual life, freedoms, memory, identity, history and rights of the Nasa. Rather than access or legal ownership, what defines territory is attachment and significance. It might be legally recognised as Indigenous territory or a broader area covering a system of interconnected 'life spaces' (such as water, rivers, mountains, etc.) or places of significance.

Communities like the Nasa who have survived colonialism, violence and armed conflict find in the territory a space to commemorate and this heightens fears of potential mining-driven territorial fragmentation or dispossession. Commemorations mark milestones in the yearly calendars of Nasa organisations, when the community revisits stories of survival and loss that mark their identity. Commemorations help internalise and recreate difficult memories (Tint, 2010). Nasa commemorations of particular conflicts involve large territorial journeys and a high degree of geographic specificity (González Rosas, 2016). Protecting the territory is protecting Indigenous memory and identity because the territory provides a space of healing remembrance. Talking about colonialism, Nazarea (2006) maintains that the severe contradictions that result from trauma call for selective hybridisation and remembering, where the territory is a source of healing capabilities to bridge the gaps of lost or interrupted memory. On the other hand, in conflict scenarios, memory reveals how the affected internalise "*lived or learned*" history, in particular as it relates to significant events (Tint, 2010, p. 240). Chosen traumas and glories unify groups and define their identities (Volkan, 2001). The past is constantly recalled and is fundamental in understanding all conflict (Tint, 2010). The Nasa define their shared identity through the territory; they have a territorialised understanding of the armed conflict. Losing or seeing their territory destroyed is a threat to that identity and knowledge.

The notion of territory, as recognised in the national and international legal frameworks, offers levers to the Nasa resistance. Rappaport (2000) and González Rosas (2016) have highlighted the legalistic elements of the Nasa resistance. The territory is a domain to assert collective rights and exercise resistance through legal avenues. The Colombian normative framework – expressed in the Constitution, laws and signed treaties – resonates with the Nasa notion of territory because it understands territory as a domain. Colombian law (Decree 2164 (1995)) defines Indigenous territories as

> *[t]he areas that an Indigenous community, group or sub-group possesses in a regular or permanent basis, and those that even though they are not possessed in such a way, constitute the traditional* ***domain*** *of their social, economic and cultural activities.*
>
> (emphasis added)

The Colombian norms include the ILO 169, the first international regulatory instrument to refer to Indigenous lands as territory (Stocks, 2005). However, the Colombian legal framework goes beyond the ILO 169 by transcending the idea of use or occupation to encompass a broader domain.

The Nasa connection with an Ancestral Territory is a basis for a range of rights, particularly concerning resource development proposals. Central among the legal levers for the Nasa moral duty to protect Ancestral Territories, in the context of extractive projects, is the requirement of free, prior and informed consent (FPIC). The ILO 169 and the UNDRIP (United Nations Declaration on the Rights of Indigenous Peoples (UNDRIP), 2007),[8] state the rights of Indigenous people to being consulted in good faith with the aim of obtaining FPIC for administrative or legislative measures that might affect them. The UNDRIP and the ILO 169 state that Indigenous people have rights over the land, territory, natural resources and environment that have traditionally belonged to them or that they have acquired, respecting traditions and customary land ownership regimes (International Labour Organisation Convention on Indigenous and Tribal Peoples, Convention No 169, 1989 (ILO 169); UNDRIP, 2007). The ILO 169 also establishes the Indigenous right to participate in the use, administration and conservation of natural resources and subsoil resources, and to being consulted should the subsoil belong to the State.

These rights provide a platform to protect Ancestral Territories as potential harm to the territory should trigger an FPIC process. The former United Nations Special Rapporteur on the situation of the human rights and fundamental freedoms of Indigenous people, James Anaya (2013), articulated the rationale for FPIC based on the risks to Indigenous rights from large-scale resource developments. Anaya (2012, 2013) stated that free, prior and informed consent should be obtained for industrial scale extractive projects in areas occupied by Indigenous people because such projects are of an invasive nature that can compromise their rights to participation and self-determination, including with regard to natural resources; property, culture, religion, non-discrimination in relation to lands, territories and natural resources; health and physical well-being (a clean and healthy environment). Because of this risk, FPIC is necessary to protect the rights of Indigenous peoples (Anaya, 2013).

Mother Earth

A third element of the Land principle is land as a sentient, motherly being, intimately close to and embedded with the Nasa, and a subject of legal rights.

Many described Mother Earth as a giver of life that nurtures and enables it. She was pictured on the walls of every Indigenous settlement or venue I visited. She appears in posters, booklets and flyers, in illustrations that turn landscapes into a giant female form. She is invoked through images of, and references to, motherhood. These portray Mother Earth as a powerful being, a force of life that gives generously in times of harmony, and brings back balance when things are out of kilter. One participant explained: "*Mother Earth gives us everything, we cannot eat without her, and we do not have water without her*", while another stated that "*we are part of her and are inseparable from her*". This bond of embeddedness with Mother Earth requires respect and reciprocity to produce a harmonious relationship. La Mariposa (pseudonym) assured me that whenever the Nasa interfere with the land, they "*do it in a considerate manner, seeking Mother Earth's permission and doing the rituals*" to sustain harmony. While an anthropocentric focus is prevalent in most environmental discourse (Mühlhäusler & Peace, 2006), the Nasa expressions about Mother Earth go further. They associate Mother Earth with the wellbeing, reciprocity and harmony that the Nasa seek to maintain, not only within the family and the community, but also harmony with Mother Earth and between all beings.

Santiago (pseudonym) told me how mining truncates reciprocity and harmony:

> *Sometimes they say that the Indigenous people are stupid; that having money [minerals] they still don't do it [mining]. But how can we go and destroy the very source of our strength, that is giving us our life.*
>
> *We are opposing mining with everything we have because for us mining is something criminal. The devastation of Mother Earth, knocking down trees, damaging waterholes, killing the fauna, the fish, and many little animals who live around mining sites – similarly to when someone forcibly displaces you from your territory – they start going somewhere else. So we reject mining and that is where we stand. We say no to the multinationals because they come to take what little is left in Mother Earth and to leave Mother Earth bleeding.*

Santiago speaks of mining as a diametrically opposed force to the life giving qualities of Mother Earth. His view reflects the sense of moral obligation to reciprocate the protection of Mother Earth.

Nasa efforts to maintain harmony and wellbeing were manifest in the Sakhelu ritual that I was invited to. The Sakhelu was in many ways about promoting healthy living as a community and a healthy territory. It celebrated locally grown food, traditional herbal medicines and the symbiosis between Mother Earth and fellow Indigenous people that results from collective agricultural work. Giving and receiving were central themes. People would exchange seeds, offer to cook, share their *chirrincho* (a traditional alcoholic drink) or play music for the others. All were expected to dance, and invited to participate in ritual food preparation and to share in the feast. The ritual included offerings such as large quantities of raw meat for the Condor spirit. The participants placed the meat on a pole, four or five metres tall. It was the trunk of a tree, cut and brought down from high in the mountains on the first day of the Sakhelu. Julian explained to me that the Elders had chosen the tree as well as its replacement and tended to them for months. This involved walking to the high altitude, remote location of the large tree to perform rituals and prepare the smaller replacement tree. According to Julian, nothing can be taken without giving something back and without performing the rituals to maintain or restore harmony. The Sakhelu provided a platform to model the considered, sharing, reciprocity relationships so important for the Nasa to fulfil their role in caring for Mother Earth, the sentient being.

The Nasa's ritual life seeks to foster respect for, and an intimate harmonious relationship with, Mother Earth. By deepening a sense of belonging and moral obligation, this compels the community to defend their territory. Julian put it thus:

> *Respect for Mother Earth is in our values. Our determination to recover our mother tongue and our traditions, such as today's Sakhelu, are acts that you could understand as religious. These are profoundly spiritual acts that express respect for seeds, for nature, for the sun, the moon, the stars, the comets. This work, through these acts, is a constant with young women and men, with our Elders, with everyone, so that we can maintain a sense of belonging and have that capacity to respond when we are called to defend this territory.*

The defence of land or territory, of economic freedom and livelihoods, of Indigenous identity and futures, so central to this Indigenous movement and to prestige within the communities, is contingent on an intimate relationship with Mother Earth as a sentient being.

While the territory is a space to assert and exercise rights, Mother Earth as a being is a subject of rights and these two sets of rights appear within the narratives of the participants as an integrated whole. El Cacique (pseudonym) linked the idea of territory as the space of rights with the idea of Mother Earth as a subject of legal rights when explaining what it means for the Nasa to consolidate an Ancestral Territory:

> *We understand the territory not as a resource to exploit but as a subject of rights, as Mother Earth, as Pacha Mama, as a living being. In an integrative manner, consistent with the cosmogony of the communities, we need to consolidate an inclusive and diverse territory, respecting each individual's right to a territory, and we need the State to respect our right to the territory so that we can continue our life as a zone [North Cauca Indigenous Zone] […] So we speak of more than the environment. In assembly, we have determined that it is life spaces that we are talking about.*

> *Mother Earth contains those life spaces: the sky, the water, land, crops. Mother Earth will guarantee the life of the Nasa people.*

The idea of Mother Earth as a being that holds rights and that protects Nasa life contributes to the legal avenues for Nasa resistance. Besides, it allows the use of legal rights to fulfil moral obligations. In linking the rights of the Nasa to their territory and the rights of Mother Earth, the Nasa are pursuing holistic guarantees for their past, present and future as a people.

In summary, in its broadest sense, Land as the basis of the Indigenous economic system, as Ancestral Territory and as Mother Earth provides the 'ground' for Nasa existence, and in abiding to the Land principle, the Indigenous people seek to guarantee a future consistent with their Indigenous identity and their ideas of a good life.

Culture

The Nasa moral principle of Culture refers to the duty to maintain and reproduce the system of meanings that supports the Nasa way of life. Hodgson (2001, p. 297) has referred to culture as "*a complex of interlocking and durable beliefs, customs and routines that permeates a group or community*". In the context of a moral economy of mining among the Nasa, the Culture principle entails the meanings and ideals such as those of Mother Earth, harmony and mutual obligation that are part of the Nasa body of knowledge, as well as the practices that embed those ideas in the community and make them enduring. The Culture principle entails maintaining and revitalising a way of life and an identity, the law, rituals, language, traditional medicine, spiritual life, forms of organisation, government and participation, music and dance that represent, embrace and protect that (desired) way of life.

In many ways, the Culture principle refers to the creative process of re-existence see p. 57) and to the resistance that has characterised the Cauca Indigenous movement. It can be argued that the separation from the campesino movement (see p. 61) effectively in the form of CRIC sought to give response to the need to construct and make visible a distinct Indigenous identity and to bring Indigenous cultural elements into the political action of the movement (González Rosas, 2016).

There are concerted, ongoing efforts to maintain the Indigenous culture, to recover it and to allow creative space for its reproduction. For example, in proximity to large towns, the Nasa have lost use of their language, NasaYuwe. The Nasa language enables the transmission and expression of Nasa knowledge. To strengthen linguistic competency, children attending Nasa educational centres receive NasaYuwe instruction. Concern about the loss of Nasa agricultural identity has seen the movement include sustainable, traditional Nasa agricultural techniques in school curricula. Rituals that were forbidden during colonial times, almost disappearing during the Republic and only kept alive clandestinely by Elders, are now being performed on a large scale, with thousands of Indigenous people and sympathisers in attendance and new elements thrown in the mix. Behind these efforts is an urge to maintain and enrich the Nasa body of knowledge as it embeds the ontological basis of their cosmogony including the economic and moral ideals that guide the predominant Nasa opposition to mining.

A 'Mining Culture' Replacing Indigenous Culture

Many Nasa associate mining with economic and power shifts and outbreaks of violence that challenge what the Culture principle stands for. Geertz (1973) referred to culture as the webs of significance in which we are suspended and that we have spun ourselves. What I heard in interviews was that as mining enters Nasa territories, culture erodes, so those webs of significance unravel. It is those webs that support Nasa ideas and ideals about, for example, nature and the reciprocal obligations within the community. Mining is seen as a powerful economic force based on the individual pursuit of short-term gains, that promotes prestige through individual consumption and accumulation rather than through community service, discipline and sacrifice. In the context of mining, the participants see customary Nasa economic values stifled.

Sek worried that miners built prestige through dangerous games of ostentation, the winner outspending the rest on whiskey, with deleterious consequences extending to women and children. This sumptuary consumption replaced contributing to working bees or serving the community through the Cabildo. Samuel (pseudonym) observed that mining facilitates a cash economy that disrupts the traditional agricultural economy by eroding the knowledge, skills and emotional attachment that sustain a commitment to agriculture, and any degree of success with it. Often, participants mentioned that young people, eager to acquire a motorcycle, a gun, a pair of expensive sneakers and the like, get involved in mining, abandoning agriculture in the hope of rapidly earning cash. The tendency to favour individual over collective wellbeing is believed to enable youth recruitment into illegal armed groups (Pequí, 2012). The frustration was evident in a car ride companion's complaint, while driving through an Indigenous reserve: "*They used to have beautiful crops in this area. Now the grass is overgrown. They only want to do mining*". Martha (pseudonym), another participant, worried that "*people do not grow their food, they begin to buy it, their diet changes …*" and in not following the traditional diet, they expose themselves to new illnesses. Lack of cultural observance was to Martha a way of risking Indigenous life itself.

Others felt that as work and consumption patterns change, distinct aspects of the Nasa culture are lost, making it even harder to recover what the movement is trying to rescue, such as language, agricultural knowledge and competence, and commitment to collective goals. Many a time I heard that when there is mining "*young people lose interest in what is traditional*". Understanding and mastery of 'what is traditional' constitutes one of the axes of prestige amongst the Nasa. It confirms a person's commitment to their

Indigenous identity and to collective causes. Stating that young people lose interest in what is traditional conveys a sense that customary forms of knowledge and understanding are losing currency among younger generations. This denotes a declining sense of belonging in those responsible for the future of the Nasa people.

The Domino Effect of Cultural Deterioration

Ultimately, concerns over the depletion of culture relate not only to its immediate signs but to the long-term effects of cultural deterioration. The visible symptoms – a teenager snubbing a traditional festival and embracing the new patron saint festivities, a patch of land filled with weeds rather than fruits and vegetables, or processed food displacing traditional foods on the table – are sad to many but are only the beginning of the story. These shifts signal a more pervasive long-term problem. Those who worry do so because they see the Nasa losing their capacity to reproduce and reinvent their way of life and to pursue their political and economic objectives. What these breaches represent is a tearing of the system of understandings the Nasa have formed through their direct observation. The obligations and ideals that moral principles embed reflect that which the Nasa know about the world. When people make choices that deviate from the cultural norm and favour working in mining, this signals their incipient abandonment of Nasa cosmogony. They could make similar choices in the future, or could lead their children to favour individual gain over customary moral principles. This means that pockets are emerging where Indigenous culture does not reproduce. Nasa culture provides the set of meanings and understandings that support opposition to mining. When mining arrives in Nasa territories, to many participants, it does not just affect the cultural practice, but it affects beliefs and ways of understanding and in doing so it threatens the economic order, because the breaches are likely to continue.

Autonomy

Autonomy refers to government by Indigenous people based on Indigenous law and culture, and to the running of affairs within Indigenous communities by Indigenous people appointed by their community. Autonomy involves self-government – government that reflects Indigenous law and principles and is not imposed from outside. Achieving recognition and realising Indigenous autonomies has been an ongoing struggle of the Indigenous movement in Colombia and the Cauca Indigenous movement has dynamised the process with proposals and implementation efforts. Indigenous Autonomy is enshrined in the Colombian Constitution, so the Nasa have autonomous management of education and health, and their own legal and administrative systems. The Colombian Constitution of 1991 recognises territorial autonomy (Political Constitution of Colombia, 1991). Therefore, cabildo authorities are in principle territorial authorities. However, subsoil property (State owned) and a lack of regulatory mechanisms to exercise that territorial autonomy mean that disputes remain. The Indigenous communities receive funds from the State and must negotiate within the mainstream government system. So, exercising the Autonomy principle requires a delicate balancing act.

The principle of Autonomy provides a basis for Indigenous resistance to resource extraction, to homogenising trends and to external projects of territorial control – such as those led by illegal armed groups. Autonomy is not only necessary for the Indigenous response to mining, but for Indigenous communities to maintain governance structures that are conducive to preserving Indigenous culture, including with regard to interactions with State institutions. Autonomy is fundamental to the interdependent network of rights of Indigenous people. Indigenous people's ability to maintain their autonomy ensures that they can choose their pathways for the future, culture and territory, and decide on the use of natural resources. Moreover, rights such as free, prior and informed consultation and consent would be difficult to guarantee without Indigenous autonomy.

Losing Autonomy to the External Powers of Mining

Those who oppose mining believe it is an obstacle to Indigenous autonomy because it creates dependency, shifts power away from Indigenous Authorities and causes environmental and social problems that force Indigenous Authorities to seek external assistance from the State to perform their functions. Many believe mining opens access for external power networks to exert control in Indigenous territories through fear and dependency, thereby threatening the standing of Traditional Indigenous Authorities. Dependency comes in the most material forms through the loss of food autonomy so losing self-reliance, becoming cash dependent and vulnerable to hunger and poverty.

Threats to Traditional Indigenous Authorities cause anxiety among the participants because they threaten community integrity. Participants feared that once respect for the Indigenous Authorities is lost, people no longer respond to calls to put the community first and to contribute as a community member. As Martha regretted, "*When there is mining, we start feeling a bit too free*". But the freedom Martha mentioned might rather be a shift of the power from the Indigenous Authorities to external powers that the participants do not trust. For example, I witnessed concerns that external money was behind mining at Munchique and that local miners were influencing the Cabildo, so that in the end the Indigenous Authorities were not governing for the local people and not enforcing the mandate prohibiting mining. Similarly, illegal armed group presence around mining, be it to deter opponents, to control mining operations or to extort payments, creates a difficult landscape for Indigenous Authorities to govern autonomously. Power dynamics shift and those with cash or weapons no longer see a reason to respect Indigenous Authorities or to work for their community.

For Indigenous Authorities, working with the mainstream State authorities to respond to mining is an exercise in demonstrating autonomy while benefitting from the external collaboration. At times the mainstream authorities might need to prosecute illegal miners, other times, such as in the murder in Munchique, they need to identify the culprits. The tensions result because such instances demonstrate that Indigenous Authorities have been unable to contain the mining problem in their territory and find themselves forced to hand over Indigenous people to mainstream authorities or to grant access to their territories. The questions people raised during the working bee against mining in Munchique exemplify these tensions. People asked, for example, "*If we are autonomous, then why must we wait for the technicians from the States Attorney to assist us? Is it that we are now required to ask for permission to stop mining in our own territory?*"

The participants believe that mining hinders Indigenous autonomy. Agricultural competency dissolves and with it so does food autonomy. As loyalties and power shift away from the Indigenous Authorities, obscure allegiances emerge that threaten not only individuals but the broader community. As a result, Indigenous leaders and communities might no longer decide what happens in their territories. Instead, those leading illegal mining, or the State authorities with the tools to deal with illegal activity or organised crime, can come to have more power or control when mining takes over.

Mining and Moral Economic Shifts

Nasa opposition to mining springs from negative associations to the moral economic shifts it triggers. Established Nasa moral principles promote behaviours that broadly align with the idea of Buen Vivir and support the Nasa organisations' economic, social, political, spiritual, cultural and environmental objectives. In more concrete and immediate terms, the principles strengthen existing networks of social, political and economic power and cooperation in Nasa communities. The participants observed an array of mining-induced transformations associated with the degradation of the Nasa world of today and with a diminishing ability to sustain and recreate it into the future. To the participants, these negative transformations result from, and encourage, miners' contraventions of Nasa moral principles.

When mining occurs within this moral framework diverse tensions emerge, revealing a duality between the Nasa economic system and the prevailing neo-liberal economic order of Colombia. Tensions arise between the collective and the individual, between home grown and imported, between long- and short-term projects and rewards, between Indigenous and mainstream culture, and between Indigenous government and external power. Those who oppose mining see miners' economic choices as selfish and damaging. In a context of limited economic opportunity, those who prioritise economic and political needs or aspirations and become miners lose moral standing in the community and their whole community loses.

Nasa organisations provide the logistical, physical, ritual, organisational, political and educational infrastructure to produce and enforce the Nasa moral principles and ideas about a good life that sustain the Nasa way of life. In his study of moral obligation, Wolfe (1989)[9] reminds us of the role of civil society organisations, such as the Nasa's, in defining moral rules and sustaining the learning process necessary to embed them. Wolfe believes that we must

> *view the construction of moral rules as a social practice, one that requires groups in civil society for the learning and development of moral rules, yet also requires that what is learned in the intimate realm be extended to people whom we will never personally know.*
>
> (1989, p. 257)

In Wolfe's study, there is a dichotomy between state- and market-driven systems of moral obligation. In the Nasa case, the organisations have built the infrastructure of moral content production precisely because they have absorbed functions from both the market and the state. They have direct access to families through educational and health services, and through agricultural commercialisation channels that allow them to connect the intimate environment of the household and the farm with the collective environments of Nasa organisations. Wolfe (1989) understands moral obligation in terms of three elements: intergenerational obligation, generalised altruism and ties of culture. The moral discourse, expressed in the principles of Unity, Land, Culture and Autonomy, and the enforcement vehicles of these Nasa organisations, from the local cabildos, to ACIN, cover all three dimensions of what Wolfe understands as moral obligation. So, significant and wide-ranging infrastructure to sustain the Nasa moral economy comes from ACIN and Nasa Authorities.

The diverse narratives of mining-induced degradation position mining as eroding the Nasa infrastructure of moral content production that supports the Nasa economic system. The concern is that mining is depleting not only people's commitment to the moral principles but the Nasa organisations and communities' ability to reproduce those principles. In addition, participants see mining as hindering possibilities to re-exist (see p. 57) or, following Albán Achinte (2009, p. 455), to dignify life by recourse to the spiritual, moral, cultural, economic, political, aesthetic and other resources and capacities of the Nasa communities. A profound fear amongst many participants is that by losing control of their own economy, they risk losing control over all other aspects of community life and losing their ability to survive as an Indigenous people. However, when survival is the main concern and where some might not be able to satisfy their needs or aspirations within the existing system, tapping into new lucrative economic activity will remain a choice.

The Nasa are concerned about the commodification of life and the influence of external economic forces that they witness with mining. Booth (1993) observed a similar concern in Marx – whose ideas have informed land rights movements in Latin America – that the transition from an economy based on "*households, their aims and the material and human means for securing them*" to an economic system "*indifferent*" to human needs resembles "*the freedom from human control for economic phenomena under capitalism*" (p. 282). However, Booth (1993) points out that Marx's idea of an embedded, household-based economy – the one the Nasa favour over the global market-driven economy of mining – is difficult to implement on a large scale without significant boundaries to the human freedoms it defends. Within Nasa communities, committed to collective land rights and economic autonomy as means to attain wellbeing and freedom, the limitations of a household-based economy are visible in that some freedoms take precedence over others. For example, the moral norm dictates that people practice agriculture and community service and not mining. However, one might ask: How does that inform and affect those who still prefer to be miners? What are their moral priorities and tensions? In the next chapter, I shall recount a set of conversations with Nasa miners to reflect on these questions.

Notes

1 Eduardo Gudynas's speech on *The War for Resources*, 40th Anniversary Session of the Latin America Group at the University of Oslo, Oslo, Norway, 14 October 2017.
2 This also reflects an overall ideological affinity not only in the rural poor of Colombia at the time, but in other Latin American countries, towards a Marxist understanding of the economy.
3 In the Colombian conflict and post-conflict contexts a number of armed actors have been present. Unlike the army, police or formal security providers, these groups have armed themselves illegally and are referred to as illegal armed actors.
4 The word that the Nasa use is 'transcend' as in going beyond in temporal and physical terms. They emphasise the idea of transcendence connecting the Nasa people across generations.
5 Some participants talked about territory, referring to something beyond the land encompassing places of spiritual and cultural significance. Territory has been the latest iteration in the Indigenous discourse and, while the term is used in all conversations, the reference to land is just as powerful and more prevalent in conversations about principles. Territory belongs more in the realm of rights-based discursive engagements.
6 The separation happened due to motivations that I shall not attempt to fully explain (or simplify) and that are not the focus of this study, but that pertain strongly to the culture principle that I discuss on p. 67. Note that the evolution of the Cauca Indigenous movement and its identity has been studied elsewhere by Rappaport (2000) and González Rosas (2016).
7 Prevalent in the south western regions of Colombia, the terraje system of work was an exploitative system that allowed land owners to profit from unpaid Indigenous labour in return for use of a small parcel of land within haciendas located in what were previously Indigenous territories (Pachón, 2005; Velasco, 2011).
8 Which Colombia has also ratified.
9 Wolfe (1989) outlines a comparative study about moral obligation in market- and state-driven systems of moral obligation.

References

Acosta, A. (2008). El Buen Vivir, una oportunidad por construir. *Debate Ecuador*, 75, 33–48.

Acosta, A. (2017). Living well: Ideas for reinventing the future. *Third World Quarterly*, 38(12), 2600–2616. doi:10.1080/01436597.2017.1375379.

Albán Achinte, A. (2009). Artistas indígenas y afrocolombianas: Entre las memorias y cosmovisiones estéticas de la resistencia. In W. Mignolo & Z. Palermo (Eds.), *Arte y estética en la encrucijada descolonial* (pp. 83–112). Ediciones Signo. Retrieved from https://moarquech.files.wordpress.com/2017/08/alban_artistasafrocolombianos_arteyestetica.pdf

Anaya, J. (2012). *Report A/HRC/21/47*. Report of the Special Rapporteur on the rights of indigenous peoples: United Nations General Assembly, Human Rights Council.

Anaya, J. (2013). *Report A/HRC/24/41, Extractive industries and indigenous peoples.* Report of the Special Rapporteur on the rights of indigenous peoples: United Nations General Assembly, Human Rights Council.

Bebbington, A. (2009). *The new extraction? Rewriting the political ecology of the Andes* (Vol. 42). NACLA Report on the Americas.

Bebbington, A., & Bury, J. (2013). Political ecologies of the subsoil. In A. Bebbington & J. Bury (Eds.), *Subterranean struggles: New dynamics of mining, oil, and gas in Latin America* (pp. 1–26). Austin: University of Texas Press.

Booth, W. J. (1993). *Households: The moral architecture of the economy.* Ithaca: Cornell University Press.

Calderón, D. (2018). Lenguas de fuego. *El País.* Retrieved from https://elpais.com/internacional/2018/09/15/colombia/1536988011_545716.html.

Caro Galvis, C., & Valencia, Y. (2012). El caso de pequeñas y medianas minerías en el Cauca: ¿Alternativas o amenazas a la autonomía indígena?. *Señas*, 2, 17–27.

Consejo de Mujeres Originarias por el Buen Vivir. (2015). *Ante-proyecto deLey del Buen Vivir.* Retrieved February, 2016, from https://groups.google.com/forum/#!topic/mesacomunicacionpacifico/gxro9YfvXos.

Consejo Regional Indigena del Cauca (CRIC). (n.d.). *Puntos de Cambio del Programa de Lucha.* Retrieved July, 2017, from www.cric-colombia.org/portal/estructura-organizativa/plataforma-de-lucha/.

Constitutional Court of Colombia. (2016a). T445–416.

Constitutional Court of Colombia. (2016b). T500–2016.

Decree 2164 of 1995, regulating law 160 (1995).

Finji, M. T., & Rojas, J. M. (1985). *Territorio, Economia y Sociedad Paez.* Cali: Universidad del Valle.

First Congress of the Black Peoples of Colombia (CNOA). (2014). *Compilación de Documentos – I Congreso Nacional del Pueblo Negro Afrocolombiano Palenquero y Raizal.* Quibdó, Colombia.

García, P. (2017). *El despojo como origen del conflicto por la tierra en el Cauca.* Retrieved December, 2017, from www.cric-colombia.org/portal/despojo-origen-del-conflicto-la-tierra-cauca/.

Geertz, C. (1973). *The interpretation of cultures: Selected essays* (Vol. 5019). New York: Basic Books.

González Rosas, A. M. (2016). *Vivimos Porque Peleamos – Una mirada desde abajo a la resistencia indígena del Cauca, Colombia.* Ciudad de México: Memorias Subalaternas.

Government of Colombia. (2010). *Plan Nacional de Desarrollo 2011–2014*. Bogotá, Colombia: Government of Colombia.

Gudynas, E. (2009). *Diez Tesis Urgentes Sobre el Nuevo Extractivismo- Contextos y demandas bajo el progresismo sudamericanoExtractivismo, política y sociedad. Quito*. Ecuador: Centro Andino de Acción Popular (CAAP) and Centro Latino Americano de Ecología Social (CLAES).

Gudynas, E. (2011). Buen vivir: Germinando alternativas al desarrollo. *America Latina en Movimiento*, 462, 1–20.

Gudynas, E. (2017). *The battle for resources*. Paper presented at the Latin America Group 40 Years, Oslo, Norway.

Guzmán Barney, Á., & Rodríguez Pizarro, A. N. (2015). *Orden Social y Conflicto Armado en el Norte del Cauca: 1990–2010*. Cali: Universidad del Valle.

Hodgson, G. M. (2001). *How economics forgot history: The problem of historical specificity in social science*. London: Routledge.

Horst, R. H. (2019). Preface. In C. V. Ødegaard & J. J. Rivera Andía (Eds.), *Indigenous life projects and extractivism: Ethnographies from South America*. Cham, Switzerland: Palgrave Macmillan. Retrieved from https://library.oapen.org/handle/20.500.12657/23130

International Labour Organisation Convention on Indigenous and Tribal Peoples, Convention No 169 (1989) (ILO 169).

Kearney, M. G., & Varese, S. (2008). Indigenous peoples: Changing identities and forms of resistance. In R. L. Harris & J. Nef (Eds.), *Capital, power, and inequality in Latin America and the Caribbean* (pp. 196–224). Lanham, MD: Rowman & Littlefield Publishing Group.

Li, F. (2015). *Unearthing conflict: Corporate mining, activism, and expertise in Peru*. Durham, NC: Duke University Press.

Llano Quintero, A. (2012). La tulpa de pensamiento de las mujeres nasa y el territorio. *Señas*, 2, 68–77.

Moore, B. (1978). *Injustice: The social bases of obedience and revolt*. White Plains, NY: M. E. Sharpe. New York: distributed by Pantheon Books.

Mühlhäusler, P., & Peace, A. (2006). Environmental discourses. *Annual Review of Anthropology*, 35(1), 457–479. doi:10.1146/annurev.anthro.35.081705.123203.

Naess, A. (2005). The deep ecological movement: Some philosophical aspects. In R. Newman, & D. Payne (Eds.), *The Palgrave environmental reader* (pp. 239–257). New York: Palgrave Macmillan.

Nazarea, V. D. (2006). Local knowledge and memory in biodiversity conservation. *Annual Review of Anthropology*, 35(1), 317–335. doi:10.1146/annurev.anthro.35.081705.123252.

Negrete, R. (2013). Derechos, minería y conflictos. Aspectos normativos. In L. J. Garay (Ed.), *Minería en Colombia: Derechos, políticas públicas y gobernanza*. Bogotá: Contraloría General de la Nación, Colombia. Retrieved from https://redjusticiaambientalcolombia.files.wordpress.com/2013/12/libro_mineria_contraloria-2013.pdf

Ødegaard, C. V., & Rivera Andía, J. J. (2019). *Indigenous life projects and extractivism: Ethnographies from South America*. Cham, Switzerland: Palgrave Macmillan.

Otero Bahamon, S. (2006). *Los conflictos de autoridad entre los indígenas y el Estado. Algunos apuntes sobre el Norte del Cauca. La integración política y económica de los indígenas a la vida de la Nación: logros, retos y resistencias*. Retrieved February, 2016, from www.institut-gouvernance.org/en/analyse/fiche-analyse-340.html.

Owen, J. R. (2009). *A history of the moral economy. Markets, custom and the philosophy of popular entitlement*. North Melbourne: Australian Scholarly Publishing.

Pachón, X. (2005). La Fuerza de la gente. Juntando recuerdos sobre la terrajería en Guambía, Colombia. Book Review. *Maguare*, 19.
Pequí, R. (2012). Guerra, proceso organizativo y juventud indígena. *Señas*, 2, 42–48.
Piedrahita, E. (2016). La tiranía de las minorías. *Semana*. www.semana.com/opinion/articulo/esteban-piedrahita-la-tirania-de-las-minorias-etnicas-en-buenaventura/481449.
Political Constitution of Colombia. (1991).
Rappaport, J. (2000). *La Politica de la Memoria*. Popayan, Colombia: Editorial Universidad del Cauca.
Rappaport, J. (2020). *Cowards don't make history: Orlando Fals Borda and the origins of participatory action research*. Durham, NC. Duke University Press.
Rico, L. (2009, July 3). *Divide y Vencerás: Nueva política uribista*. Retrieved from https://lasillavacia.com/historia/divide-y-venceras-nueva-politica-uribista-2803.
Rodríguez-Franco, D. (2014). The rise of popular consultations. *The Americas Quarterly, Spring 2014* (Consulta Previa and Investment). Retrieved from www.americasquarterly.org/content/rise-popular-consultations.
Scott, J. C. (1977). *The moral economy of the peasant: Rebellion and subsistence in Southeast Asia*: New Haven: Yale University Press.
Scott, J. C. (1985). *Weapons of the weak: Everyday forms of peasant resistance*. New Haven and London: Yale University Press.
Stocks, A. (2005). Too much for too few: Problems of indigenous land rights in Latin America. *Annual Review of Anthropology, 34*(1), 85–104. doi:10.1146/annurev.anthro.33.070203.143844.
Taussig, M. T. (2010). *The devil and commodity fetishism in South America (2)*. Chapel Hill, NC: University of North Carolina Press.
Tilly, C. (2005). *Identities, boundaries, and social ties*. Boulder, Colo.: Paradigm Publishers.
Tint, B. (2010). History, memory, and intractable conflict. *Conflict Resolution Quarterly*, 27(3), 239–256. doi:10.1002/crq.258.
United Nations Declaration on the Rights of Indigenous Peoples (UNDRIP). (2007). 61/295 C.F.R.
Universidad Autonoma Intercultural Indígena (UAIIN-CRIC). (n.d.). La Universidad. Retrieved August, 2022, from https://uaiinpebi-cric.edu.co/la-universidad/#resena.
Velasco, M. (2011). Contested territoriality: Ethnic challenges to Colombia's territorial regimes. *Bulletin of Latin American Research*, 30(2), 213–228. doi:10.1111/j.1470-9856.2010.00500.x.
Volkan, V. D. (2001). Transgenerational transmissions and chosen traumas: An aspect of large-group identity. *Group Analysis*, 34(1), 79–97. doi:10.1177/05333160122077730.
Warnaars, X. S., & Bebbington, A. (2014). Negotiable differences? Conflicts over mining and development in South East Ecuador. In E. Gilberthorpe & G. Hilson (Eds.), *Natural resource extraction and indigenous livelihoods: Development challenges in an era of globalization* (pp. 109–128). Surrey, Burlington: Ashgate Publishing Ltd.
Wolfe, A. (1989). *Whose keeper?: Social science and moral obligation*. Berkeley: University of California Press.

4 Alternative Perspectives on Mining

Introduction

The presence of Nasa miners in Nasa territories demonstrates the practical and moral tensions, difficulties, dilemmas and conflicts putting the mandate of a NO to mining in Nasa territories to the test. Positions and actions on mining are not unitary or undifferentiated. Different collective or personal realities have led to diverse responses, including the decision to engage with mining, or at least not to oppose it fully. Beyond the prevalent NO response, there is nuance and tension when it comes to responding to miners that are members of the Nasa community. Such cases provide an opportunity to learn about how Indigenous Authorities and individuals make decisions or adopt responses to mining and invite reflection about the practical and technical aspects of the NO mandate. Some Cabildos, such as Delicias, have chosen to conduct mining in their territories to keep foreign miners at bay while minimising the financial costs of doing so, and localising the benefits of mining. Other Cabildos, having done mining before, were winding this activity down, mostly on financial grounds; they judged it economically unviable, based on prospective incomes and regulatory compliance costs. Some community members had become full-time, part-time, seasonal or occasional miners. They maintained that through mining income they had been, or would be, able to invest in housing, their farms or their children's education, in an economic environment where such opportunities were hard to come by. Mining researchers or analysts believed that the NO position was a political statement, lacking technical backing and analysis. These different collective and individual choices across different territories demonstrate the complexity of mining questions for the Nasa of North Cauca as a collective. I shall visit these positions through the explanations the participants offered, while reflecting on the key tensions or dilemmas these examples highlight. These participants wanted to see the NO mandate further analysed. In light of their own circumstances or research experience, they either oppose, or seek to see more nuance in, the mandate. When embarking in mining, they face moral contradictions that are not dissimilar to those faced by rural Latin American populations when engaging with the capitalist economy.

DOI: 10.4324/9781003226895-4

There are difficult material conditions that inform these alternative positions about mining. Statistics throughout this book are limited, because during 2016 and 2017 Colombia was yet to undertake a national census with appropriate differentiation of its Ethnic Peoples and because the most recent statistics at this time were from 2005. Nevertheless, some information on economic circumstances was available based on ACIN's census of Indigenous cabildos from 2008 and 2009. The available data sketched the outlines of a precarious material situation for the Indigenous people of North Cauca. The population of Resguardos was predominantly Nasa (95 per cent) and young, with over 58 per cent under 29 years of age, including 28 per cent younger than 15. Twenty-five per cent of residents were students. School attendance was lower than for other groups (except after 30 years of age) and Indigenous people had lower literacy than other groups (ACIN, 2011). The economy was mostly agricultural (37 per cent of residents in addition to 16 per cent who were seasonal agricultural workers) (ACIN, 2011). However, Resguardos were, on average, tight on land with less than a hectare per person for agriculture, housing, recreation and ritual purposes, and some, like Canoas or Guadalito, were crowded. So, Nasa Resguardos were unable to produce sufficient food for the dietary needs of their population (see National Department of Statistics, 2005). Although National Government data can only be used for illustrative purposes[1], it resonates with the Nasa statistics in indicating that significant proportions of the Nasa populations were experiencing hunger[2] and had lower educational outcomes than other groups (National Department of Statistics, 2005). These statistics describe communities that were facing a hunger problem and scarcity of land, despite their predominantly agricultural vocation. They had low levels of schooling and a large seasonal workforce. It is within this overall landscape of material poverty and limited financial stability that we need to contextualise Nasa responses to mining and the alternative positions I describe in this chapter.

I shall begin by examining the narratives of four Nasa people who are involved in mining ventures or work as miners. One of them comes from a Resguardo where mining is allowed. The remaining three are from Resguardos that adhere to the NO to mining position. I shall outline some of the motivations, feelings, frustrations and attachments of these participants in relation to mining. Later, I analyse the perspective of a Nasa environmental and economic expert with first-hand experience in a Nasa region that has gradually phased out construction materials mining due to financial and technical limitations, and that has grappled with the more complex dynamics of informal alluvial gold mining. This discussion will provide insight into the tensions that communities with limited livelihood options and capital face when dealing with mining. Later, I shall examine the positions of a group of Nasa who are not miners but researchers or analysts and did not fully oppose mining. They called for more analysis to inform the debate on mining. They believed the Nasa needed more technical data on social and environmental matters so that the debate can be better informed and the position on mining can reflect moral principles as well as current realities. I shall conclude by discussing the convergences and contrasts between these positions.

Forced into "*Socially Driven Mining*"?

Some Indigenous Authorities undertake mining in their territories to prevent external miner entry. One such example was the Delicias Resguardo. A participant from this Resguardo, Diego (pseudonym), shared with me a moral discourse in support of the Delicias community's kind of mining. Diego's rationale for mining gold at Delicias encompassed: i) a lack of alternatives to defend the territory from foreign mining other than the Indigenous community practising mining or assuming significant financial costs for surface rent, ii) invoking elements of ancestrality through collaboration with the local Afro-descendant community, who have traditionally practised mining, iii) reinforcing and giving continuity to a relationship of mutual collaboration and support with the local Afro-descendant community, and iv) efforts to implement what roughly translates as "*a socially driven mining*" (*una minería social*). To Diego, these elements make mining the lesser evil compared to external miner entry that would likely result in unfair outcomes for the community. Diego described the result as a locally legitimate, involuntary mining that seeks to mitigate social impacts and deliver local benefits.

Diego explained that gold mining was not the community's first choice, but a consequence of the regulatory framework that left them with no alternative (see also Caro Galvis & Valencia, 2012). The Colombian legal instrument of Indigenous Mining Zones allowed Indigenous Authorities to apply for preferential rights for mining concessions (Law 685 (2001), Art. 124). Several Nasa Resguardos filed these requests and there were two Indigenous Mining Zones in North Cauca Indigenous territories at the time of the study. Once the Indigenous Mining Zones were established, it became clear to the Indigenous Authorities that to maintain this preferential right they would have to pay surface rent. Not paying would open the door to others to apply for mining concessions. So, preventing mining company access to Indigenous territories came at a cost to the Cabildo once an Indigenous Mining Zone was established. Interviews with some leaders from ACIN and from at least three North Cauca Cabildos revealed that these Indigenous Authorities were unaware of surface rent costs before applying to establish Indigenous Mining Zones.

Indigenous Mining Zones have been found to create a propensity in Resguardos to shift from agricultural production to mining (Caro Galvis & Valencia, 2012). Diego explained that within the economic circumstances of the Delicias community, the surface rent payment was a significant amount that was unsustainable for the Indigenous Authority, so they decided to mine the gold, in collaboration with the nearby Afro-descendant community. Diego recalled that after years of paying surface rent it totalled about COP $300 million (or over USD $99,000), a considerable amount for the community. This community is made up of mostly rural workers and employment is not always available. As an illustration, a research participant estimated that a day's wage for a seasonal rural worker in the region would be around COP $20,000 (USD $6.61) at the time of the study. The magnitude of the surface rent payment required to protect the territory from external miners was considered unfair.

The collaboration with the Afro-descendant community was an essential, legitimising aspect in Diego's discourse on Delicias's mining. Diego thought that this collaboration brought an element of ancestrality, community mining credentials and continuity for the strong relationship between the Afro-descendant and Nasa communities. Diego described mining in Delicias as a collaboration with the Afro-descendant community in the area, who had "*done mining for 200 years*". Ancestral mining enjoys much wider community acceptance than other forms of mining in the region, because it is a significant component of Afro-descendant livelihoods and culture (Weitzner, 2012). With this history of Afro-descendant community mining also came mining expertise. Nevertheless, analysis performed by ACIN's former think tank, *Casa de Pensamiento*, has described the mining enterprise at Delicias as medium-scale rather than ancestral mining and described the mining cooperative as an alliance between the Cabildo and two individuals belonging to the Afro-descendant Community Council of Mazamorrero (Caro Galvis & Valencia, 2012).

The joint mining venture is an expression of the communities' ongoing relationship of collaboration and mutuality that Diego illustrated with several examples. In all these stories, resistance against external threats to the communities and their territories was a defining aspect of the interethnic relationship. Several research participants from Delicias and beyond described collaborations with their Afro-descendant neighbours in resisting violent groups seeking to forcibly displace them. In the Delicias area, during the 2000s there was a strong paramilitary incursion and the paramilitaries forcibly displaced the Afro-descendant community of the village of Mirasoles. Diego described how the Nasa community of Delicias hosted some 60 families in the Resguardo's school for a month, providing shelter, food and education; how later they went as a large group of around 1,000 people to talk to the commander of the paramilitary group and asked that they vacate the area; and how it was the Indigenous community, not the Government, Diego emphasised, that arranged the Afro-descendant return to Mirasoles. It was also Mirasoles that hosted the mining operation. At the time of our conversation, the Afro-descendant community did not allow access to the mining area to anyone external to the Afro-descendant or Indigenous communities, explained Diego. The two communities have a history of protecting and supporting each other. Their mining partnership, seen as resistance to foreign mining, is a continuation of their collaborative, interethnic resistance relationship of mutual support and reciprocity.

Setting aside the fact that it involves practising mining, Delicias's strategy of cooperation with its Afro-descendant neighbours resonates with the increasing emphasis in interethnic and intercultural collaboration in Indigenous mobilisation strategies. Intercommunity collaboration has emerged as a new source of strength and a favoured strategy for North Cauca community organisations. Interviewees from various organisations spoke of collaborations between campesino, Afro-descendant and Indigenous communities as necessary to realise collective rights, address social exclusion and deal with conflicts and threats from violent actors. Participants consistently spoke of the merits of interethnic collaboration. Almost invariably, Indigenous participants would acknowledge that mining is important culturally and economically to some Afro-descendant communities, demonstrating the mutual respect necessary to collaborate. Examples of these collaborations include dialogue committees to discuss shared problems, and joint social mobilisations, including actions taken against illegal miners. A frequently mentioned example was when higher numbers of Indigenous Guardia were needed to expel illegal miners from a Nasa territory and the *Cimarrón* (Afro-descendant) Guard joined in the hundreds, making the action a success.

What did Diego mean by "*socially driven mining*"? Mining with localised benefits, socially inclusive outcomes and minimal negative social consequences. Rather than individual, short-term wealth, socially driven mining should result in long-term community and family benefits. Diego emphasised that in Delicias, "*the miners are not rich like in other places*". Rather, the Cabildo invests mining revenues in community projects such as housing for elderly leaders and guides community miners to use their mining income for child schooling and housing improvement. In contrast, people who work in mining ventures not managed by the Cabildo, stated Diego, do not prioritise education or housing. But Diego was convinced that the "*socially driven*" nature of mining at Delicias made it possible to mitigate some of the problems observed in mining areas nearby. For example, he described a "*healthy social environment*", free of substance abuse and prostitution, where wealth is not concentrated but distributed fairly. This was in stark contrast to what he and other participants (Indigenous and non-Indigenous) evaluated as a harsh and unhealthy atmosphere in nearby localities with external miner presence where "*they pay them money, but then on Fridays they open the bottle shop and by Monday people have no money. They open brothels as well, food prices go up, and it is all a vicious cycle*".

But where does Mother Earth stand in this discourse? Diego accepted that mining damages the territory and told me about plans for projects on "*clean mining*" to reduce or phase out chemical use. At the same time, Diego thought that the argument of not doing mining to protect Mother Earth was not valid because others were already practicing mining. Rather, he thought it was best if the community did the mining so that they could localise the benefits and do it more responsibly. Diego understood mining by Indigenous people as territorial protection and a guarantee of good management. In contrast, when it came to mining by external parties in Indigenous territories, his view converged with the prevalent official position of NO to mining.

Delicias illustrates how under specific legal settings, and with the perceived threat of external miner entry, Nasa communities with limited financial resources can devise alternative moral discourses about mining. In this case, the discourse does not deviate completely from the prevailing moral framework and discourse on mining, rather it draws on some of its elements to arrive at a different conclusion: that mining by Indigenous people can be a form of territorial defence that can be carried out with the Afro-descendant community, when there is a threat of foreign mining entry and the financial costs of preventing it are prohibitive for the community. This line of thought incorporates elements of the search for Autonomy, the protection of Culture and Territory, and Unity in facing external threats. Paradoxically, by diverging from the movement's stance on mining, this position also poses tensions around the Unity principle. Most importantly, addressing chemical use was an intention but there was no clear timeline on this at the time of my fieldwork. This raises questions about possible environmental legacies on the Delicias territory or beyond and their implications in light of the Territory or Land principle. Some participants from other Indigenous territories voiced such concerns.

Delicias has taken an autonomous position in direct tension with the Indigenous organisations it is part of, while strengthening the relationship of mutuality with the Afro-descendant community and consolidating an interethnic resistance to external mining. "*The relationship with ACIN when it comes to mining has been very difficult*", Diego told me. Despite this tension, Delicias remains active within ACIN and CRIC. Delicias people have led key networks at ACIN, and held Councillor seats in both ACIN and CRIC in recent years, including during 2016, 2017 and 2018. However, this does not preclude the Delicias Authorities from exercising the Autonomy principle. La Mariposa had explained to me that only Cabildos with their communities could make decisions about mining based on their local circumstances. Delicias has decided to operate a mine, while strengthening the resistance collaboration with their Afro-descendant neighbours.

Mining Livelihoods in NO Territories

> *We should look not for the negative but for the positive side in each thing […] A miner already said that gold has no enemies. He only has friends. He helps you. He is like a friend. So, lets look at this, lets promote some leaders so that mining is more organised, let's show this to people by talking.*
>
> (Botón de Oro, Indigenous miner)

Although most Nasa Cabildos affiliated to ACIN held an official position of NO to mining in their territories, there were still Indigenous miners within those territories. What led these Indigenous miners to choose a livelihood that their Cabildos had banned? What were their moral economic perspectives on mining? These are the questions that I address in the coming pages, by analysing the experiences and views that three Indigenous miners from NO territories shared with me. Here, I shall refer to them with the pseudonyms Botón de Oro, El Águila and Buena Vista. In their narratives we shall observe that the themes of resistance, fairness, moral tensions and material limitations emerge once again as difficulties in responding to mining, but take very different expressions in a NO to mining territory.

Botón de Oro, El Águila and Buena Vista were consistent in describing a range of material circumstances that lead people to become miners. All of them felt that people, in particular those aged between 20 and 40, turn to mining because they lack economic opportunity. Agriculture and pastoralism, the favoured livelihood alternatives within the prevailing Nasa moral framework, are not sufficient to make a living according to these participants. Local market conditions for agricultural produce are harsh. For example, Botón de Oro told me "*It is very hard to work the land and to go down to the town with your starchy foods and have to come back with all the produce again, having sold nothing*". They might have limited or unproductive land, or no land at all. The Cabildo states it is unable to provide funds to support economic projects for miners. But Botón de Oro and El Águila contend that politics defines who gets assistance and the miners are left out. All three admit Cabildo funds are limited and some expect that those limited funds be managed differently although, being earmarked by the Central Government, there is limited room for this. Nevertheless, they feel isolated and disappointed in the Cabildo. Overall, most grievances pointed to structural problems about local livelihoods and to political power.

Botón de Oro, El Águila and Buena Vista did not see themselves as part of the mining-connected violence or environmental problems. Instead, other miners, who use chemicals or who refuse to share with others, were to blame. They dismissed the environmental complaints of their neighbours: they took adequate care, they did not use chemicals, the effects were negligible and complaints might mask envy rather than tangible environmental damage. The violence within the community as a result of mining caused great concern among these participants. They believed the murder of the miner in Munchique was born of greed. The shared principle among these three miners appeared to be that Mother Earth gives gold for all and it cannot be hoarded. Instead, everyone should be welcome in a mine. So, the participants did not feel part of these violent dynamics. They were characteristic of a different approach to mining that did not reflect their principles.

"Gold Has No Enemies, He Only Has Friends!"

Botón de Oro grew up seeing people in his family mining and did not see a moral contradiction in being an Indigenous miner. As a child, he would bathe in the river while a relative panned for gold to later "*sell and turn into money*". A second relative also panned for gold, and used to pawn it to obtain cash. So, in his early teens, following family tensions, Botón de Oro tried his luck with mining, away from home. Inexperienced and still a child, he did not succeed. However, later in life, he tried his luck again and found gold in his home territory.

From then on, gold opened a path into independence and freedom for Botón de Oro. Mining provided an opportunity for economic self-reliance that, despite the risks which he acknowledges, is to him superior to other economic activities. He is unwilling to work for the minimum wage because that would make him feel exploited. So, mining offers Botón de Oro a way to work independently and achieve higher earnings, without the servitude he fears. Botón de Oro told me:

> *The others say to me that I should do something else, use my talents better. But when you show your talent, people exploit you. Why would I want to go and work around the clock for someone else for the minimum wage? I'll stay with the gold. I can pace myself. I know that if I push myself, I do it for myself, that if I lose, it's my loss.*

To Botón de Oro, mining comes with higher risks and also with higher returns. He told me how days and weeks can go by without finding gold until one day compensates the losses of weeks. He estimates that a good mining day can yield more than three weeks of minimum wages. So, mining offers a much better livelihood alternative than working for the minimum wage.

Botón de Oro described gold as a mysterious, magical, unpredictable, but ultimately reliable and fair ally. He told me of gold's fairness: "*If nothing comes out on the day, then nothing comes out on the day, but gold will compensate you. He compensates you for the time you lost. He's like a piggy bank*". Botón de Oro also spoke at length about gold's unpredictability and about unexplainable things that had happened to him as a gold miner. He told me, still in disbelief, how years back, he panned for gold in a creek for several months but found nothing. Then, someone else spotted a kernel of gold and decided, despite everyone's advice, to pan the same creek for gold, making significant earnings on the same day. Botón de Oro still finds this surprising and told me that he and others associate such events with magic and with the sprits of Mother Earth, such as El Duende:

> *Nowadays people let that creek swell and they wash it again, it's like something magic. […] I come from the Indigenous side, so people start telling stories. Is it that El Duende throws out the gold for our sustenance? Because, you see, often I would come back there in the morning and wouldn't find my tools and then I'd come back and find them there. I didn't know whether someone hid them.*

Botón de Oro told me that "*people believe that Mother Earth has spirits, El Duende is the one among them who gives permission to extract gold and, if you ask, he gives*". But Botón de Oro did not ask for permission from El Duende.

He did have his own rituals for when he started working. He mentioned some rituals like lighting a cigarette that he said he was not sure why he practiced. Others, like cutting a lemon in half and doing a cross-shaped incision in it, to later place it at the entrance of his place of work, had a specific purpose, in this case to "*absorb the negative comments*" of his critics. Botón de Oro instilled his responses to the mystery surrounding gold with a similar degree of mystery. It is as if he refused to submit to any powers, asserting his own control through his personal rituals, without engaging in rituals for the spirit who controls gold.

These examples of fetishism reveal the grievances of rural peoples over an unfair economic system that does not reward their efforts and that offers no care. There are parallels between this way of understanding a substance, gold, as an entity, and the forms of fetishism Taussig described in the neighbouring area of southern Cauca Valley's sugar cane plantations and in the Bolivian tin mines that Nash studied (Nash, 1993; Taussig, 2010). Taussig describes how peasants or Indigenous people, when exposed to proletarian work, resorted to the idea of the devil to mediate their relationship with money and, through magic rituals, conveyed the exploitative and destructive nature they saw in capitalist relationships. However, Botón de Oro attributed to gold the qualities he did not find in capitalist market relations. His participation in the market had been frustrating. No earnings came from his agricultural work and he considered the pay from other salaried work unfair. So, Botón de Oro devoted himself to the search for gold and in his narratives presented gold almost as an employer, who offers the rewards and fairness Botón de Oro had not found elsewhere in the market. He referred to gold as an entity with agency, a being that acquires obligations to compensate the work of the miners, and an entity that is fair. Investing time panning for gold, Botón de Oro placed his future in the hands of this unpredictable, but benevolent, entity. In both cases, a form of fetishism is at play. In one, the destructive characteristics of capitalist relationships become the substance of the fetish, whereas in the other, it is what the market cannot offer that the miner finds in the benevolent gold.

But being a miner is not all about rewards; people criticised Botón de Oro for being a miner. He read this as envy, disregard for his economic circumstances or lack of knowledge about how much foreign mining was already taking place where Indigenous people did not benefit. Botón de Oro believed it was people's envy that led them to say he was "*destroying Mother Earth and polluting the water*". Underneath people's suggestions and complaints, he saw demands for him to make a less generous living, and this was presumably the basis of his belief that moral and environmental grievances masked envy. Botón de Oro found people's complaints that he is damaging Mother Earth surprising, because of the chasm he observed between the damage his livelihood created and the damage that other mining activity caused. He saw the NO position as a hindrance to an appropriate response to mining in Indigenous territories. Botón de Oro was worried that others would reap the financial rewards of mining a finite mineral, while Indigenous people prevent their fellow community members from doing so and the Colombian Government mining institutions favour corporate over community miners.

Botón de Oro deemed the Cabildo response to local Indigenous miners wanting, because it tried to end an activity that was difficult to bring to a halt, rather than focusing on improving the way people practiced mining and on ensuring the community derived some benefits. He believed entire hamlets could obtain a livelihood from mining if the mineral deposits in the area were well mined. Frustrated, Botón de Oro told me:

> *I hired an engineer to get me a permit and he told me that this mountain is already sold, sold to a mining company. So, who will support the Indigenous people? The Government can throw the army at us. Many people are fighting and we are fighting over the crumbs. If what they tell me is true, this mountain might have raw gold that you can cut with a saw. If they see there is a problem, then they should help us to do it better. There is no way to stop mining anymore. But it is not being done properly, the gold is left in there because of bad extraction.*

Botón de Oro believed violent mining-related conflicts among Indigenous people should not happen if people really subscribe to Indigenous understandings of Mother Earth. He said mining was pointless if there was going to be death and suffering. Instead, according to Botón de Oro and others, some Indigenous miners were seeking to prevent others from entering mines and "*they shoot to kill […] and have tried to involve the guerrillas asking them to plant landmines to protect access, but the guerrillas will not engage in that*". Botón de Oro and others assured me that many former guerrillas were making a living from mining. Interest in practicing mining by Indigenous and external parties, together with the material needs of returning FARC members, had convinced Botón de Oro that "*mining can no longer be stopped*" in post-agreement Nasa territories.

"I Do it Because I Have No Alternative, but I Am Ready to Stop"

Buena Vista (pseudonym) is another member of the Nasa community who works in mining in a territory where the Cabildo opposes mining. I interviewed Buena Vista at the home of one of his relatives, in his Resguardo where we were allowed privacy in an open area to have a conversation. It took a while to get to meet him and for him to be comfortable to talk. On the day of the interview, he was very keen to offer his views as long as he was not identified. So, I am taking extra precautions in the amount of context I provide to prevent identification.

Buena Vista combined mining and agriculture to make a living for his family. Mining income had enabled investment in his farm and improvements in his family's standard of living. He was fond of his mining work. The first thing he said to me was:

> *I must tell you that [mining] is something very beautiful, something I like a lot. It is risky, it is dangerous, but it becomes part of you, you feel it and you miss it, not all the time, but you never forget about it.*

The possibility of creating "*value*" or wealth for him and his family emboldened Buena Vista, in particular because he found very few other alternatives. His mining work consisted of buying material out of the mining tunnels to process it in a mill to extract gold. Buena Vista told me that when he looked at the soil he purchased, he felt encouraged to think that it had value and from there would come resources to invest in his farm or housing for his family. Paralleling what has been observed in several countries in Sub-Saharan Africa (Chigumira, 2018; Okoh & Hilson, 2011), this complementarity of artisanal and small-scale mining, and agriculture contests the idea that artisanal and small-scale mining erodes agriculture.

"Not Everything is Pink and Rosy"

A more nuanced understanding of mining surfaced when I asked Buena Vista what would happen if the younger people in his family practiced mining and when we spoke about the recent mining-related assassination. With future generations in mind, Buena Vista started speaking of mining as a last resort for him and expressed more concern about the negative effects of mining. He was definitive that he would not encourage future generations to mine because of safety, environmental and moral reasons. "*We ought to stop*" he said, "*I would tell them it's very dangerous. A rock can fall at any time and leave you with a disability or take your life*", he added. More so, he said that water was more important than money to the wellbeing of future generations and that they should find other forms of earning a living. Later, speaking about the violent death of a young miner at Munchique, Buena Vista expressed deep disappointment and concern, and stated that mining must stop. "*Not everything is pink and rosy*", he said in a sad tone, "*things are getting out of hand, it is not the armed groups, it was the community itself – these things shouldn't happen*".

Buena Vista was concerned that instead of establishing a dialogue, the miners and the Cabildo have confrontations. He thought the Cabildo were making decisions without talking to the local miners, thus damaging relationships and delaying a solution. I found scant evidence of any dialogue. He emphasised that the problem with mining was not a local but a national one and that in his territory many people aged 15 and over, both men and women, were involved in mining. With this he seemed to invite a much deeper reflection about what motivates Indigenous miners and about how to create new livelihood alternatives.

In the context of the Peace Agreement implementation, questions about local livelihoods became more complex for Buena Vista. While he observed increasing levels of crime that he attributed to FARC demobilisation, he also feared that the post-conflict would bring renewed military efforts to stop miners, who do not have titles or environmental licences. These fears relate directly to the question of livelihoods in a security environment that does not offer sufficient guarantees to Indigenous workers. The fear of punitive measures may make mining livelihoods less feasible locally and, in a context where economic alternatives are limited, crime can hinder both mining and agricultural livelihoods.

"Here and all Over the World, Mining Will Not Stop"

El Águila had worked in the mining tunnels since the second half of 2000. I went to his residence to interview him and he was very open to speaking yet conscious of the wording of his responses because of the contentious nature of his livelihood in the Resguardo. He was open in sharing with me the reasons people work in mining. El Águila was more hesitant when it came to discussing the relationship between mining, conflict and peace. I believe he did not want this study to apportion blame for local violence. Like Buena Vista, El Águila described mining and agriculture as complementary livelihoods for many people in the local Indigenous community. Through his work, El Águila told me he was in contact with over 40 people, mostly young, who work a coffee and a mining season each year. From around November to March, most of his fellow miners went to tend to the coffee and only four remained at the mine. In the off-season, from June to September, the number of miners would swell back to over 40 people, he explained.

El Águila emphasised that mining work was very tough and offered no safety nets, only monetary benefits. I asked what he liked about mining and he replied: "*What is there to like? I like the money I can earn, nothing else, I would rather work in agriculture but I have no land*". The risks are significant. The miners dig the tunnels with picks and shovels, and install wooden structures to prevent cave-ins. They enter the tunnels without any safety equipment besides a torch. El Águila joked that "*mining is for machos*". They spend up to seven hours engaged in heavy physical work, either with the pick or carrying buckets full of rock and soil out of a 50-metre-long tunnel. There are no assurances or contingencies. When I asked whether he considered himself a mining leader he promptly clarified that the responsibility was individual:

> *No. Each person is independent. Everyone is free. When we go into the tunnel, I tell them: you come at your own risk and you should tell your mum and your dad and your children that you come in here and you don't know if you'll come out. Nobody gets paid here, you are free to come and go.*

El Águila was sceptical of the argument that local miners harm Mother Earth and must be stopped. He thought while mining happens in other Indigenous territories or elsewhere it does not make sense to stop him and his fellow workers. El Águila associated efforts to control mining with ACIN more than with the local Cabildo. "*Why do ACIN allow mining in [other Resguardos]?*" he asked rhetorically, "*If they allow it there, they should allow it here*", he argued. However, neither the Cabildo in his locality nor ACIN have jurisdiction in other Indigenous territories. ACIN can only act at the request of a local Cabildo. If a Cabildo authorises mining, ACIN is not allowed to interfere uninvited. Nevertheless, El Águila's discontent reveals a disapproval of Indigenous resistance to mining on the grounds that mining is happening in other places in any case. "*In South Africa, they really are taking out the heart of the Earth*" he told me, so (he implied) why not resist that? For the local Indigenous Authorities and zone organisation to coherently oppose mining, El Águila believed their opposition should extend to mining anywhere in the world. One might say that in multiple ways it already does.

For El Águila, far worse things were happening in Indigenous territories that should have concerned people more than mining. He believed that ACIN and CRIC should concentrate on controlling the larger evil of drug trafficking, a far more brutal trade based on his experience as a coca grower. To him there was a hierarchy of morally undesirable activities, where drug trafficking was far worse than mining, and needed to be controlled first. He believed people and Cabildos from territories where the drug trade was active are in a tenuous position to campaign against mining on moral grounds.

El Águila drew out several connections between mining, violence, the armed conflict and the post-conflict, although the first thing he said to me when I interviewed him was that mining was not in any way connected. He discussed the activities of FARC's Sixth Front in the area and their interaction with mining. The Sixth Front had many Indigenous militants. El Águila told me that many left FARC around 2008 and returned home, where in absence of other work opportunities "*thank God, they began working in mining, and left [FARC]*". He saw mining as a way of keeping people away from armed group militancy. At the same time, he believed since 2014 life was much better in the area and the miners had been able to see the fruits of their work, as opposed to during the intervening period from the mid-2000s, when the Sixth Front was very active attempting to recruit young people. But as far as mining was concerned, El Águila believed the Peace Agreement would not change things. "*I might stop, but others will continue*" and the Peace Agreement would not make a difference because "[miners] *are already working regardless of whether there's peace*", he added.

Several common elements are palpable in the narratives of the miners from the NO territories and even that of Diego whose Resguardo had a legal mine. First, mining goes far beyond the Indigenous territories, and people see it as something that cannot be stopped. Second, financial or material considerations rank high in decisions about participating in mining: either the miner has no other sufficient livelihood alternatives due to lack of land and unfavourable microeconomic dynamics in local markets, or the Cabildo has no financial resources to cover the surface rent. Third, the miners see their work as a form of resistance against external mining capturing the benefits. In all, there is a wave of economic opportunity that the miners feel they have no option but to ride. However, in other scenarios, where people are not invested in mining to such an extent, there are still those who question the NO position and want to discuss it in light of recent experiences and data. I shall examine some examples of these narratives next.

When the Moral NO to Mining Makes Financial Sense

Luis (pseudonym) is an economic-environmental expert from the Nasa community, who spoke with me in detail about the multiple faces of mining in Resguardos of the Toribío municipality. There had been multiple legal mining enterprises by locals and outsiders as well as waves of informal activity and these made Luis a uniquely positioned analyst of mining dynamics in Nasa territories. These mining experiences include marble extraction for spiritual purposes (coca leaf chewing), external commercial marble mining, Indigenous mining enterprise attempts (construction materials) and informal alluvial gold mining. To Luis this conversation was "*an effort to adopt a critical point of view, a self-critical perspective, dispassionate and neutral*". In our discussion, Luis analysed in some detail a range of opposing perspectives: the "*political positioning*" behind the NO to mining, the financial precursors of informal alluvial gold mining together with its deleterious effects on rivers, the failed attempts at formal Indigenous mining of construction materials, and the moral concerns over the prospects of predatory large-scale mining. Luis contended that the NO to mining was a political position to protect the territories, but acknowledged that a lack of livelihood alternatives for local families pushed them into informal mining. His view was that the political and moral positioning missed the underlying problem – that families in need oscillate between illicit economies that are detrimental to communities and territories. Many armed confrontations between FARC and the Colombian army happened in Toribío, as did many creative community peacebuilding activities. Luis is well versed in analysing armed conflict dynamics and explaining them to outsiders. In doing so, he was clear that, in the Resguardos of this municipality, the bulk of the violence was drug-traffic rather than mining-related. In this context, Luis was optimistic about the prospects of the Peace Agreement for the Indigenous communities.

The story Luis told me is one of Resguardos where mining has been legally and illegally present, but where the markets at the time of the interview did not favour the activity, making it relatively painless, at least for the time, for local responses to mining to converge with the NO position. He spoke of financial dilemmas and moral tensions about mining. Experiences with external and with Indigenous-run mining enterprises presented the first financial and moral dilemma. Lacking the capital and technical support, Indigenous miners could not sustain a profitable business for their community and meet legal requirements. Luis described the Indigenous Authorities' frustrated aspirations thus:

> *The idea was to create an enterprise, the Cabildo is like our big enterprise that represents all this community territory. It's our great institution. The workers do the labour, and we incur the production costs of any economic activity in the Resguardo, let's say the mining process, the profits go to a community fund that has to support other community projects. The mine is not the Cabildo's. It does not belong to the miners. It belongs to the entire Indigenous community, because it is in an Indigenous territory. However, that dream did not come through. Why? Because at the time the miners didn't have the capital, didn't have appropriate machinery – one digger is not enough – they didn't receive technical advice, they didn't have professional personnel. So, this left them doing a very artisanal exploitation well below the requirements of a profit generating business. So, they were never able to transfer a profit to the Cabildo. Now the miners are old. They are asking who will pay them for the work and the truth is that there are no funds for that because they didn't save. By now, there's been significant progress to fulfil the legal requirements, but still the Cabildo is seriously considering closing those mines and declaring them sacred sites. But there is somewhat of a magnet there, because of the sheer amount of construction materials. We did a study […] there is enough material to work for at least 100 years. But again, meeting the legal requirements means machinery, capital and technical advice.*
>
> […]
>
> *Besides, they had done the market studies during the narcos boom in Cali, when there was a lot of demand for marble for ostentatious home facades. That came to an end too. So, they couldn't reach the targets. Now, 20 years later, let's say the machinery is of another standard now.*

There are commonalities between the aspirations that Luis outlined and those that guide mining at Delicias – a type of mining that is oriented to support communities in the long-term. However, while Delicias continues, somewhat protected by the favourable gold markets, it has been more difficult in the Toribío municipality. Overall, the experience with several mining ventures in these Resguardos has not been encouraging. Tacueyó, Toribío and San Francisco, the Resguardos of the Toribío municipality, held several mining environmental licences and established an Indigenous Mining Zone, as Delicias did, with the same stated intent: blocking external mining interests. Indigenous Mining Zones require the holder to conduct mining, but having a profitable and compliant venture has not been easy without capital and technical capacity and given the markets for construction materials. For instance, the Tacueyó Resguardo had four environmental licences, three of which had been cancelled on compliance grounds, and a fourth was similarly nearing cancellation in late 2016 or early 2017.

An Indigenous mining enterprise seemed unviable, but the alternative, letting an outsider with financial capacity extract the local resources, seemed to Luis like opening the communities and territories to external predators. He commented:

> *The position is to always oppose large scale extraction, never to allow multinationals to basically deplete what little is left of natural resources in Colombia. Although we know there are mining concessions in various places in this country [...] here in North Cauca but also in the Amazon. You feel a great sadness because it is like the last stage, finishing off this historic pillaging process they have put Colombia through.*

The first element of the moral tension in Luis's narrative transpires as such: Indigenous miners' attempts at a morally acceptable mining failed because of lack of funds. Outsiders, on the other hand, had the funds, but could not be trusted to act in a morally sound way. According to Luis, a seemingly "*easy*" way to resolve this tension for Indigenous Authorities was the NO to mining. But he believed the NO was a political position to block "*the multinationals*". If mining cannot be done for the benefit of the Indigenous people, looking after Mother Earth and within the requirements of national law, then others should not extract local minerals only to leave the Indigenous people more impoverished and stuck in a polluted environment. However, Luis was concerned that the NO overlooked the livelihoods question, so important to Indigenous families. He contended that allowing what sometimes is wrongly labelled artisanal mining in Indigenous territories, on the basis that Indigenous people have practiced it, was a weak argument on environmental grounds. Luis explained to me that:

> *There is a position of NO to mining, but I think of it as a very political position, mostly of the leaders. It is very easy for a leader, who has a secure livelihood, to say NO to mining. It is very different for a community member, living in an Indigenous territory, with no other income besides mining. At times they have explained the NO to mining as a strategy to preclude multinational entry. But if you look deeper, there's been mining in Indigenous territories. So, they say artisanal mining is admissible. But artisanal mining is as damaging as large-scale mining!*

However, the concern is not just environmental. Material poverty, a more structural problem driving families into informal mining and crops of illicit use, remains unresolved. Luis told me that alluvial gold mining came to the area in the late 2000s. Attractive profits motivated people in need who, in pursuit of higher margins, used harsher technologies like dredging the rivers. Luis said:

> *I want to tell you about a type of mining whose echo really resounded here: alluvial gold mining. The rivers of these three Indigenous Reserves don't*

> *have a lot of gold, but there is still some left and there might be veins somewhere and people who pan the river for gold can find some gold there. That can easily give a family one or two grams. In today's market, that's $160,000 COP [over $50 USD]. This was intense in 2009, 2010, 2011, the prices, the Indigenous people intervening the rivers. Then if they offer Indigenous people some technical improvements so that they don't get two but ten grams of gold, that's when the famous dredges make an entrance. First, you have this simple tool, the pan; then the sluice; and then the dredge and that one comes with a water pump and a whole suit of things (sighs)… Don't get me started.*

Then, in what Luis called a paradox, the highly profitable illegal marijuana crops displaced alluvial gold mining and "*saved the rivers*" but begun corroding families, with the youngest first in line. So, a section of the local Indigenous community becomes trapped in a swinging motion between the socially damaging drug trade and environmentally damaging alluvial gold mining. Frustrated, Luis said:

> *From 2011 onwards, marijuana crops started to appear. Thanks to the marijuana crops, today we are not looking at dry riverbeds. But what happens when the marijuana crops come to an end? At the beginning prices were extremely favourable and they are still very favourable. But when that's over, the rivers will be the victims. So, it's a paradox. There is a political positioning of NO to mining, but there is a need at the territory level, they don't say anything, they just do their mining.*

The presence of illegal economies such as crops of illicit use and illegal mining has overlapped in the Toribío municipality adding complexity. The level of armed conflict that the people from Toribío endured was extreme. The Tacueyó area, located in Toribío, is known in Colombia for the 1985 Tacueyó Massacre or Roberto Franco Massacre, when FARC's Roberto Franco's front guerrillas executed over 160 of their own over fears that they were National Army moles or informants. Toribío also saw numerous guerrilla-army combats. The battlefield was often inside the human settlements. For example, community members dismantled army trenches built within the urban area of the municipality. Illegal armed groups also engage in extortive activity. Crops of illicit use and illegal mining can be susceptible to illegal armed group extortion. So, I asked Luis about the interactions between these livelihoods, and the parties to the armed conflict, in the context of the Peace Agreement. Luis had observed one period where mining connected directly with the conflict in Toribío. For a time, FARC mined gold in the mountaintops. Then the army expelled them. At the time of our conversation, Luis estimated "*90 per cent of the armed conflict violence is drug traffic related […] because each pound of marijuana attracts a $10,000 COP [roughly $3.3 USD at the time] extortion payment*". Regretfully, he added "*this is very difficult, I mean, the* *armed conflict exists because we Indigenous people contribute the fuel for it*". Where in other Resguardos, inhabitants linked mining to the armed conflict, based on instances of Army-FARC combat over a mining deposit, to Luis the same situation was insufficient to draw a link in Toribío. Presumably, what is in evidence is a different sensitivity to precursors of combat, and perhaps also an indicator of combat proximity to human settlements, because in this case the combat was on the mountaintops of the Toribío municipality.

As for the Peace Agreement's interaction with mining, Luis was positive. He observed that "*there are no risks, because there is a strong legal shield protecting Indigenous territories that multinationals will have trouble breaking*". Luis believed the Peace Agreement was, in general terms, a source of opportunities, because it would help the Nasa Indigenous community free itself from the stigma of having FARC in their territories. Many people had labelled them as guerrilla supporters or militants because of that.

Given the risky nature of marihuana and alluvial mining economies, I inquired about other livelihood options. Luis assured me that the alternatives were there, but the problem was scaling them up to cover enough families. He gave me some examples and explained the limitations and the resulting problem:

> *We have the trout alternative, the dairy alternative, the fruit juice alternative, the coffee alternative [...] But first, they are not enough to cover the needs of all the population and second, drug trafficking creates an economy of easy money. You don't need much of a technical procedure to plant marijuana as you would for granadilla or gulupa [tropical fruits], or top quality coffee. So that is why I say this is tearing apart the social fabric, because it is creating an artificial economy. If it continues, we will hardly be able to bring back the youngsters of this generation to grow food.*

But if the livelihood situation renders the NO to mining contradictory, there is still another moral argument in support of the NO: Indigenous people did not extract minerals for profit, but for ritual purposes; commercial mining came on the back of outsiders and it did not align with Indigenous moral principles because of its effects on Mother Earth. According to Luis, marble extraction was a traditional activity. Indigenous people would use the dust for coca leaf chewing (*mambeo*), a male spiritual practice. Luis described the nature of traditional marble extraction like this:

> *The guidance from our Elders is that we should use the land respectfully. Mining was always used here as an instrument, it was only to extract the marble to get a powder, called mambe, used to chew the coca leaf. That was the real use of mining. That was based on a very important value for us: barter. So, the Indigenous people who needed mambe had no problem in bringing a load of vegetables and starchy foods to give in exchange for some marble rocks from the mine. That was the value. But when the*

> *colonisers arrived [...] the Indigenous people also went down that market path and begun exploiting the marble for commercial purposes in the 1960s and 1970s.*

Commercial exploitation was not in line with Indigenous values. According to Luis the "*position is not to do mining, not to do mining to be coherent with our principles*". This was Luis's starting and concluding point. He also explained there was some discontent in the community about the environmental legacies of construction material mining because "*there is a fair bit of inert material left*". While those legacies did not lead to community conflict, they might have compromised the miners' political and moral standing within the community.

Luis's analysis and narrative come from a very informed, deeply embedded perspective and a profound understanding of the workings of the Indigenous organisations at the zone, regional and local level. Although he saw the NO to mining as a political position, Luis was deeply concerned about what alluvial mining does to the territory and, in some respects, concurred with the NO position. He offered an empathetic account of the material and market mechanisms that see Indigenous people involved in alluvial mining and unable to mine construction materials profitably, within the parameters of the law; together with a critical opinion of the NO position, in all seeking to remain impartial within a tense debate. Overall, Luis's analysis reflects the following logic:

- i Indigenous leaders oppose mining in Indigenous territories because:
  - a Indigenous people have been unable to practice mining for communal benefit and within legal requirements
  - b External people will not do mining for communal benefit, but will devastate the territory and leave the locals to deal with the damage
  - c Mining damages Mother Earth, clashing with Indigenous principles
  - d In the past, Indigenous people only mined for ritual purposes
- ii The NO to mining overlooks the material conditions of Indigenous communities so leaders adjust the argument:
  - a There are insufficient livelihood alternatives for local families and this makes alluvial mining attractive
  - b So-called artisanal mining is considered admissible because Indigenous people have practiced it
- iii The so-called artisanal mining is as damaging as large-scale mining and this renders the adjusted position contradictory
- iv Both the lack of livelihood alternatives and alluvial mining's deleterious consequences on the territories stand, creating an ongoing economic-moral tension that had, temporarily, shifted to another damaging market, that of marijuana

v Ultimately, mining contravenes moral principles and it was only part of Indigenous identity as part of ritual life
vi To be coherent Indigenous people must not practice commercial mining.

Through analysis of the available markets and environmental effects of both large-scale construction materials and alluvial gold mining, Luis arrived at the conclusion that the reasons in point i) and the adjustment in ii)b) are contradictory or invalid in light of the Nasa moral principles and an ancestral past that holds a particular moral value. Rappaport's (2000) work on the memory of the Nasa explained how Nasa morality integrates strong references to an idealised past that leaders and community members recall in order to provide moral validity and continuity to present actions.

Luis's reflections are also valuable as a description of a community learning process on mining, framed in changing market, legal, socio-political and environmental circumstances. Gradually leaders and community members have adjusted their position based on livelihood and political priorities. Diverging positions result from different material situations and varying levels of political visibility. This raises questions about the political and moral costs that Indigenous miners might face. We have heard some of this from the miners earlier on in this chapter. But there is more to examine. For example, we have not yet looked into whether and how the material power or wealth that results from mining interacts with the Indigenous political sphere. The testimonies in the next section include some reflections about that question, as does Chapter 6.

The learning process that Tacueyó and other Toribío Resguardos have gone through with mining might be a reflection of the processes that were taking or needed to take place at the movement level. I next look into the experiences and reasons that have led some Nasa to call for more research and technical analysis to inform the Nasa debate on mining. These participants were calling for a movement wide learning process about mining integrating not only the moral and livelihoods debate, but a more considerate process of data gathering and analysis to inform the dialogue.

A 'Technical' Discussion: Does Evidence Support the NO?

Some Nasa participants with substantial and diverse knowledge of mining called for a more technical and pragmatic discussion to complement the debate on morality and livelihoods. They shared the view that the NO was a political strategy and that, as important as the moral framework is, it is necessary to also build a link between Indigenous policies on mining and a socio-environmental analysis of mining. They called this a "*technical*" discussion. I shall examine the arguments of three participants who arrived at this conclusion through active participation in research or dialogue processes on mining. I refer to them with the pseudonyms Cxayuce, Sek and Andrés. They all had research credentials and moral and intellectual standing in the

Nasa community. That there was no "*technical argument*" behind the NO position was a concern for the three of them, but for different reasons. Sek and Cxayuce believed there were deeper social, political and economic processes at stake beyond the livelihood question. Andrés hoped there would be ways that Indigenous people could enter beneficial partnerships with reliable mining companies or investors.

Mining, Indigenous Authorities and Indigenous Political Contests

Sek and Cxayuce were concerned that, in the Canoas Resguardo specifically, a lack of research, information and dialogue about mining trapped the community in an ongoing dispute. This dispute, according to Sek, was spilling into the political arena. To the argument that mining in Indigenous territories is "*a purely economic problem*", Sek responded with detailed descriptions of the multiple layers of deterioration that mining brought to Canoas (discussed in Chapter 6). I shall first address Sek and Cxayuce's descriptions of the political disputes surrounding mining. Sek spoke of the significant political weight that water quality and availability carry in Canoas and of a more recent political currency of the mining-livelihood question. Sek described a contest between these two forces that, to reach resolution, needed better information. In absence of this, the disputes would turn into challenges to Cabildo Authorities in Canoas, Sek feared. This aligns with the views of the mining research group of Canoas (Caro Galvis & Valencia, 2012). Cxayuce regretted the limited willingness to support research and to participate in dialogue about mining in Canoas that contributed to perpetuate and escalate internal community conflicts.

In Canoas, like in Tacueyó, there have been different types of mining or mining interests: colonial era gold mining, present day alluvial and underground gold mining, multinational miner entry attempts, and mining of construction materials for local infrastructure, where the community accepted the latter as legitimate. Colonial hacendados begun to mine gold in the Canoas area, using an enslaved African workforce (Caro Galvis & Valencia, 2012). Sek had visited some of these very old mining tunnels that had been closed. His account was that after closing the tunnels:

> *They began agriculture, with coffee, then when the Cabildo and community recovered the land there was livestock. Then, in 2008, the mining boom affected Canoas. In 2009, there was contact with Anglo Gold Ashanti for a consultation, but it wasn't very formal, it was just talking.*

During that gold mining boom, there were both underground and alluvial mining in Canoas, and both were relatively under control at the time of the interview according to Sek. Alluvial miners were using dredges and the community expelled them. But that was not the end of it. Sek told me, "*The problem is that our own community started doing it, so it is very tough to expel them*". According to Sek, at the peak of mining activity, an approximate 250 Canoas families were directly involved in mining, almost 10 per cent of the Canoas population. However, several forces converged to bring it down to what Sek estimated was about 20 per cent of its peak levels. First, on three occasions the community gathered in large numbers, sometimes as high as 2,000 people, to block the mining tunnels on the sacred mountain of Cerro Munchique. Second, the Cabildo announced it would take back collectively owned Cabildo lands if the users were practicing or authorising mining on those lands. Third, there were tense discussions within the Cabildo, where people with strong environmental concerns, mainly over water, asserted they would not allow mining under any circumstances. The final element facilitating a decline in mining was that, according to Cxayuce, many Canoas families have private land in Munchique and once Canoas Authorities tightened the grip, mining shifted to Munchique.

Sek and Cxayuce were worried that previous Cabildos had been unwilling to collect data on mining dynamics, leading to a stagnant debate and escalating conflict. During the gold boom of the late 2000s and in the late 2010s, mining had been a divisive issue and the Cabildo had lacked information to feed into any potential discussion. When asked about the scale of the mining workforce in Canoas, Sek responded:

> *There was a proposal to do a survey of miners and a questionnaire ready and the Cabildo thought it was not viable. [...] Instead, they asked in the Assembly: who supports mining in Canoas? So, 50 per cent raised their hands. Who doesn't? And 50 per cent again. So, they couldn't resolve it. There was a lot of mining at that time, so many families went to the Assembly and they couldn't resolve it. The Cabildo didn't have the knowledge about this matter to put forward arguments for a debate. [...] The Cabildo said: it is a double edge sword, because even those who are not miners can end up filling out the survey. So, maybe it was going to be too many people. So, the Cabildo thought they would lose.*

Also holding the Cabildo back was fear of collecting information on an illegal activity. Sek told me:

> *The Cabildo never supported getting the data on how many murders and suicides came from mining. I think that was out of fear, because around that time there was a 4WD that was carrying the gold and there were five people transporting it and two of them were murdered. We got to that level. They waited for them to kill them and take the gold from them. We didn't find out who was behind it, who was financing it, etc. [...] It becomes research about an illegal activity. Of course, the Cabildo was afraid of doing that.*

To Sek, there was at least one missed opportunity in not doing the research because mining tunnels are located in the vicinity of water intakes and that information would have strengthened the position of the NO side.

A further obstacle to informed dialogue, for Sek, was a lack of political finesse amongst some members of the Cabildo. He said, "*The Cabildo hadn't learned that you have to be very political to talk to people who are full of anger and hatred, and in the Cabildo, we can see that there are people who are so radicalised with mining*". I asked what radicalised meant. Sek explained, "*People who would not allow mining no matter what, so they cannot even come into the discussion, you know, I listen to the other's argument and put mine forward. No, they are like 'No, means no'*". I enquired what motivated those positions. "*Those are the people who started the environmental movement in Canoas, and they talk about the water springs, and the forest, and it is a very valid argument, and I think that is a factor in the decline of mining activity in Canoas*", Sek responded. So, this environmental position has helped reduce mining but has also silenced other positions, hindering dialogue and ultimately the chances of an agreement or some kind of shared solution.

Cxayuce had similar concerns and added that a crucial obstacle to resolving the mining problems in Canoas had been an overall lack of Cabildo willingness to participate in dialogue exercises that had been prepared explicitly to discuss the problems with mining. Cxayuce explained:

> *At the beginning, in several occasions, there were attempts at dialogue and the miners would come but the Cabildo would not show up. So, I want to highlight that. I think that is something that didn't help. So, the miners say 'but we go and the Cabildo doesn't'. I think it has to do with that same lack of tools and knowledge about mining.*

Dialogue had been absent and instead there had been direct confrontations, like the ones at the mining sites. These, according to Sek, had come close to becoming physical. Sek's experience had been that the spiritual work the Traditional Doctors performed when the Canoas community implemented mass activities to halt mining was what helped prevent violent confrontations. Sek told me that during one of the actions the miners stood by the tunnels with machetes, while at the other, in a mining mill, the miners had firearms. But Sek was relieved that "*people were serious and there was no violence*".

Miners continued to gather power as a political force that could influence the Cabildo make up through votes, because the Cabildo Authorities are popularly elected at the Assembly. For Sek, mining had become a vehicle for political contestation and it was necessary to understand the reasons. The views and feelings we heard from some of the miners earlier on appear in some respects to confirm this concern. Some of the miners resented being passed over in Cabildo land distributions, felt relegated and unrecognised for their contribution to community life, felt their work was not valued, or found themselves at a loss looking for alternative livelihoods and felt there was no empathy towards them. Without attempting to draw a straight causal effect between political grievance and engagement in a morally contentious economic activity, it is nevertheless necessary to acknowledge a palpable political tension that would be far less feasible were it not for i) the economic power that mining grants its participants, in particular those higher up in the production chain, and ii) the co-dependency networks of teams working in mining tunnels. Sek and others (see Chapter 6) observed that as far as political effects go, mining resembles drug trafficking in redistributing power.

The main concern for Sek and Cxayuce was that, devoid of information for a debate, the community continued to go through mining-related confrontations rather than informed and measured dialogue. Those holding a firm NO position were having difficulties connecting the moral argument with first-hand data, while miners had been open to discussing the disagreement, but for how much longer? Sek and Cxayuce wanted to see research playing a strong role in Nasa mining policy-making and dialogue. These testimonies illustrated the view of some participants that, without studying what is happening, it is not possible to establish positions and that regardless of moral standing, the miners' grievances and reasons needed to be heard. The views of Sek and Cxayuce speak of the lack of "*data*" in the mining contentions, and suggest fears and scepticism in the NO and YES sides.

Technical Debate and a New Kind of Mining

Andrés has been involved in various analysis exercises about mining and was certain that a technical analysis also needed a space within the Indigenous movement so that moral and political positions could be better supported and more nuanced. He assured me that there was yet another position on mining within the Indigenous movement – the possibility to say yes to mining, but not at any price, only under certain stringent conditions. Andrés, like Luis, Cxayuce and Sek, saw the absolute NO to mining as a political position that was not based on "*technical*" analysis. He believed that the Indigenous movement must build compelling technical arguments about its positions on mining. In Andrés's words:

> *There is a position that is the absolute NO to mining. That is more of a political position. It is not technical. It has not been planned or thought out. But I believe the Indigenous movement will have to start thinking this through, because all Indigenous peoples in the world have used mining, but not the ways extraction is done today. That debate must happen. Find whether there are ways to reconcile the traditional experience of many Indigenous peoples with the very advanced technological experience of the mining companies of the world.*

A central concern for Andrés was that, by implementing imported mining models and aspirations that do not respond to Colombia's specific fragilities, the country might turn its mineral endowment into a curse. This is how Andrés explained it:

> *I believe Colombia has a very particular situation, because when you talk about mining in Colombia you must talk about the entire ecosystem it has.*

> *It is one thing to talk about mining in Chile and Peru, and it is another thing to talk about mining in Colombia, or mining in Africa or in other places where the conditions are not the same. I believe it is worth discussing this, because maybe we have a privilege in Colombia, but it could also turn into a curse, a tragedy if we don't look after it. So, it is about thinking that through. The Indigenous movement must be constructive and seek to compel. Now the world is talking about bio-civilisation, because of global warming. Here we are talking about harmony and equilibrium and, in a way, the approach is philosophical, in terms of dreams, so we need to make it something concrete and that is the debate within the Indigenous movement.*

Andrés saw an important role for Indigenous knowledge in the Colombian mining debate. This was an urgent debate for Andrés who believed Government capacity to negotiate with mining companies was limited. To cover that gap, he thought multiple sectors should come together and hold a technical debate to define the directions and conditions for mining. Within that debate, he saw a crucial role for Indigenous knowledge. He stated:

> *The Indigenous movement must prepare itself for a major technical discussion. If you want to contribute to building something, you must offer something you can debate. That is where we have a shortcoming, but it can be overcome. But a dialogue has to start between private companies, who have their own respectable perspective; the environmentalists also have their perspective and it is very valid and respectable; the Indigenous movement has a perspective and it is respectable; and the economists have their own models and their own reasons. Just to mention those four. Before talking about a licence or an exploration permit, the first thing is to open a space to define the policies, the mechanisms, the vision, the rules. It is very important to think about this here in Colombia because I have just recently come to realise that the Government doesn't really have the capacity to negotiate with private companies.*
>
> [...]
>
> *So, this really is a challenge for the Indigenous movement, to prepare itself to generate dialogues. Not to give in. But to negotiate about rules and conditions that are important to protect biodiversity and ecosystems.*

Andrés thought that to be heard, Indigenous arguments must come packaged as technical arguments. For example, he explained:

> *The Indigenous person might say 'no mining here, this is a sacred site'; the other person might tell you 'no mining here because it is a source of water and it will destroy the source of water'. So, we need to explain the arguments in technical terms.*

Andrés was certain that the Indigenous contribution could be part of productive collaborations with even the most unexpected sectors of society. For example, he recalled that at one point, concerned about an external proposal to mine in a *páramo* area (a subalpine environment with high conservation value), the Indigenous movement contacted the wealthy industrialists of Valle del Cauca (the neighbouring region) and exposed the potential deleterious consequences on water sources. Andrés explained that:

> *The Uribe Government had granted a licence for a study for exploration and exploitation and the Indigenous movement organised a march and approached the rich people in Valle del Cauca and told them 'look, there is this whole problem here, and the only solution here is to intervene in the area and to suspend the licence'. They made an alliance with a completely antagonistic sector, because for the rich people of Valle, Indigenous people are guerrilla people. But based on the reasoning of the Indigenous people, they did a big technical exercise with the CVC [Valle's Natural Resource Management Authority]; they brought the best biologists, geologists, and they made a study. They said, here is an argument from the Indigenous people, we are going to make it technical and argue: No! Suspend the licence, this is not a problem only for the Indigenous people, this is a problem for the future of Valle del Cauca. You know what I mean? That is when I say that we have to talk about values and about knowledge.*

Andrés, Cxayuce and Sek advocated for a more informed debate and for dialogue, in some cases internally within the Indigenous communities and, in other cases, with other actors. Their priority was to ensure that the debate about values comes also with data to support the arguments, such as the claims about environmental damage, and that the data compares mining with other economic activity that also has environmental repercussions such as agriculture. There are two central elements to this call for a technical debate: one is about testing the moral discourse with hard evidence. For example, did the argument that Indigenous people should make a living from agriculture stand a thorough test? Or: Did the miners really lack other alternatives? On the other hand, there was a sentiment that moral principles needed to be translated into technical terms so that they did not appear merely ideological and could be explained in a manner that others understood. With this came a confidence that environmental facts supported Indigenous moral statements. There was also some evidence of openness, if only incipient and cautious, to talking with mining companies, although the weight of the discussions people held with me indicated that this view was the exception rather than the rule. Overall, where participants had experience in research and analysis related to mining, they had become strong advocates of embedding research processes within the Indigenous movement political decision-making and advocacy efforts.

Discussion

The narratives in this chapter have a number of points of convergence. First, the participants faced contradictions between the tangible dynamics that surround mining in Indigenous territories and the requirements of the prevalent value system of the Nasa community. As a result, most of these participants were after a reality check, a pragmatic test of the NO position in the context of material circumstances, and economic, legal and environmental dynamics. After that test, some are favourable to mining, find no choice other than to do it, or remain unconvinced and curious and would like the debate to go further. The choice to embark on mining is not straightforward. Instead, it is marred with moral contradictions that participants have devised mechanisms to resolve, responding to individual and collective circumstances. Once people become miners, tensions and contradictions aside, the next question is how to do it. The answer gravitates around fairness. Ideas of fairness influence who can do mining and who it should benefit. It is the idea of fairness that determines mining's success either at the collective or the individual level. However, contradictions do not only surface within individual narratives, but across them. This highlights all the uncertainty that surrounds mining and the need for more localised information to illuminate what most see as a well overdue debate.

In various ways, these participants were asking for a debate about the movement's position on mining because this position must reflect what the Nasa know about the world and, as far as mining is concerned, the world is changing rapidly. Can and should the Indigenous people oppose mining? Can they oppose mining when others do not? Does it make sense to oppose mining while it advances in other places? Is it fair to ban people from what they contend is their only livelihood? The participants spoke about the material, legal and economic limitations that Indigenous communities and individuals face when dealing with mining. In some cases, they concluded there is no other option than to practice mining. Maybe agriculture does not offer sufficient earnings, or there is insufficient land. Maybe there is no other way to protect the territory from foreign miners because the legal framework is such. Foreign miners will do it and the Government will favour them, so why should one not do it first? When so many companies are doing mining, why be concerned about Mother Earth? It is impossible to stop mining, so why try? In other cases, the participants were less certain. Instead, they called for more analysis and for openness to the insights it might offer.

Once individuals or communities decided to participate in mining, the moral tensions and dilemmas continued both at the movement and personal level. To several, the NO position created tensions and problems as it overlooked the precarious material circumstances of those with least economic choice, who participated in the illegal economies of alluvial gold mining and marijuana cultivation. Individual miners faced other types of tensions that came from their transgressions and limited economic options. They felt shame, they saw themselves as the targets of envy and some felt unfairly blamed for environmental damage. However, there was insufficient data to discuss environmental damage. They saw mining as out of control, unstoppable, something more powerful than themselves or anyone else. They acknowledged the violence, said they would stop, but no sooner had they said it than they would start to hope they could continue. Some disliked mining and did it only for the money. Others thought mining was something beautiful, but swiftly painted a different picture for fear their descendants might take up the dangerous occupation. All, however, saw mining money as positive and constructive, not barren as most of the other Indigenous participants of this research did, but as a source of capital to invest in their farms, housing and family. To the moral arguments, one of those same torn miners responded with accusations that some leaders condoned drug trafficking and unfairly attacked mining. However, there is evidence that mining had come to take an important place in illegal groups' finance portfolios (Rettberg & Ortiz-Riomalo, 2016). While this was clear to many participants, it appears the miners were less willing to acknowledge that connection. The tensions grew and bred further conflict.

Miners resorted to diverse strategies to resolve the moral tensions of their economic activity. These strategies hold similarities with those that Nash (1993) described in her study of Bolivian tin miners. Nash (1993) described the tensions miners faced as a result of: exploitation and high dependency on their mining work, of the contrast between home and work value systems, and of the process of proletarianisation. The way some of the participants in this chapter resolved, at least in the interim, their own moral tensions resembled some of the strategies that Nash describes, namely, compartmentalisation in time and space, creating a hierarchy that prioritises one moral value over another, and merging one's interests with those of new others. When, as agriculturalists, the participants were not able to afford Buen Vivir, they prioritised survival or even 'living better' over the value of Unity. By compartmentalising in time and space, miners could see their livelihood as acceptable in the now, but not in the future and not for their loved ones; they could condemn violence and commit to abandon mining, while asserting that mining was beautiful because it improved their families' standard of living. When fellow Indigenous Cabildos do not support the decision of a Cabildo to mine, other alliances, such as with Afro-descendant communities, have taken precedence – thus interests are merged with those of new others. Similarly, if the rest of the Resguardo condemns mining, loyalties to other miners can become stronger than those to other members of the Resguardo.

Next is the question of how to do mining. Regardless of their financial success, my concern here is how these Indigenous organisations or individuals believe mining should be. In those deliberations, the Indigenous participants in this section embraced ideas of fairness. If mining happens in Indigenous territories, it should be for the benefit of the Indigenous community, said proponents of collective mining; or for the benefit of the Indigenous people, said the individual entrepreneurs who saw mining rewards as fair and mining work as a dignified way to free themselves from exploitation and to prosper.

The fact that others had the capital to invest in mining did not give them the right to take the resources from Indigenous territories. If Indigenous people extract minerals from their own territory, it should be for the benefit of all. For example, Delicias and Tacueyó participants asserted that a mine should be the community's, not solely the miners'. Moneys that come from mining should go into community projects and capital goods rather than sumptuary consumption. Or, in the case of individual miners, the profits should be for capital investment that will benefit families in the long-term. In this, the participants shared the fear of barren mining moneys – those that do not contribute to the life of the family or the community. This fear parallels the commodity fetishism and devil beliefs Taussig (2010) documented in Colombia. So, Delicias and Tacueyó decided to guide expenditure away from that risk, while individual miners emphasise its avoidance as a principle guiding their behaviour.

The question of capital, or lack of it, surfaces time and time again, highlighting the central role that material limitations have in shaping responses to mining. Lack of capital is a barrier to Indigenous mining ventures that seek to be profitable while operating within regulatory parameters. It is a barrier to alternative livelihoods that might sway Indigenous people away from informal or illegal mining. Without capital, it was hard to make mining viable in Tacueyó. Delicias was still at it but had yet to address its environmental legacies and for that it would also need capital. In parallel, lack of investment in agricultural projects to a scale that can offer viable sources of income to compete with mining was a recurrent complaint for miners.

In this context of material scarcity, mining provides a vehicle for moral, economic and political contestation. For the miners from NO territories in this chapter, mining money is as available as the Cabildo money is not. These participants have hoped and continued to hope that the Cabildo would invest in them by giving them land or through agricultural projects. Some felt rejected, unacknowledged and disadvantaged because they had not received land, or recognition for their contribution to community life. Through mining they have found an avenue for independent advancement. Setting the urge to avoid employment and legal liabilities aside, there are people higher up in the mining production chain who have undeniable power over other miners. By 2018, conversations with people from the same Indigenous Reserve reveal that mining supporters were exerting higher levels of influence in the Cabildo through elected representatives. Mining as a redistributor of power has enabled or intensified these contestations. In parallel, co-dependency and camaraderie relationships have developed amongst the miners, a group of people who have found no satisfactory response to their economic grievances from their Indigenous Authority. Mining is disrupting the socio-economic order and, with its contradictions, also placing the moral order under tension. But the experience with mining is so diverse that amongst the Nasa we also find examples of mining strengthening political alliances within the Indigenous community and with the Afro-descendant community, like in Delicias, an alliance that Diego believes has helped keep more damaging forms of mining at bay.

In many ways this discussion is about a moral economic crisis. A contradiction in the economic system of the Nasa, and in its interaction with other economic systems, was testing how the prevailing moral principles were applied. Those testing conditions came on the back of desires for a better life and the idea of fairness. Through contested economic practices, Indigenous communities or individuals were pushing moral interpretation towards favouring economic opportunity and the improving living conditions that it promises. There is a certain angst to pursue the opportunity to mine that comes from a conviction that minerals belong with Indigenous people first. This is implicit in the ideas of current and former Indigenous miner participants.

There are many contrasting and contradicting narratives and one wonders whether, aside from the grievances, the needs, the perceptions and tensions, the balance of other evidence tilts either way; in other words, does anyone have an overall understanding of what is happening with mining in these Indigenous territories? Does anyone know the number of workers and their material conditions, the volume mined, the amount of water used, the chemical discharges, the status of water quality, the links with armed and other illicit groups, the extent of ancestral mining, the degree of Indigenous involvement, the volumes that remain in the soil and rivers? These are the next obvious questions. Questions that to Cxayuce, Sek and Andrés needed answers. For them, only with the data will it be possible to have a constructive debate both within the Indigenous communities and outwards. A conversation was overdue, but there were barriers to discussing mining other than in a moral monologue. What some participants called for, and many others seemed to need, was an intercultural analysis exercise to provide material for a debate that is moral as much as it is technical. These participants were seeking a link between intercultural science and research, and the moral framework on mining.

Notes

1 What the Nasa refer to as North Cauca, the National Government post-2005 census studies split into two zones that do not cover all municipalities. One set of statistics agglomerates Caldono, Corinto, Jambaló and Toribío, while the other aggregates Santander de Quilichao, Suarez, Buenos Aires, Miranda and Caloto (National Department of Statistics, 2005).

2 The fasting index measures the percentage of people who, due to lack of money, were unable to eat any of their three main meals during one or more days over the course of the week prior to the census. In the first zone, the fasting index was at total of 15 per cent and in the other 16.8 per cent. Overall, people in urban areas fare slightly better than rural dwellers (National Department of Statistics, 2005).

References

ACIN. (2011). *Plan territorial cultural.* Santander de Quilichao, Colombia: ACIN.

Caro Galvis, C., & Valencia, Y. (2012). El caso de pequeñas y medianas minerías en el Cauca: ¿Alternativas o amenazas a la autonomía indígena?. *Señas*, 2, 17–27.

Chigumira, E. (2018). Political ecology of agrarian transformation: The nexus of mining and agriculture in Sanyati District, Zimbabwe. *Journal of Rural Studies*, 61, 265.

Law 685. (2001). Mining Code.

Nash, J. (1993). *We eat the mines and the mines eat us: Dependency and exploitation in Bolivian tin mines.* New York: Columbia University Press.

National Department of Statistics. (2005). *Censo Nacional*. Bogotá, Colombia: DANE.

Okoh, G., & Hilson, G. (2011). Poverty and livelihood diversification: Exploring the linkages between smallholder farming and artisanal mining in rural Ghana. *Journal of International Development*, 23(8), 1100–1114. doi:10.1002/jid.1834.

Rappaport, J. (2000). *La Politica de la Memoria*. Popayan, Colombia: Editorial Universidad del Cauca.

Rettberg, A., & Ortiz-Riomalo, J. F. (2016). Golden opportunity, or a new twist on the resource-conflict relationship: Links between the drug trade and illegal gold mining in Colombia. *World Development*, 84, 82–96. doi:10.1016/j.worlddev.2016.03.020.

Taussig, M. T. (2010). *The devil and commodity fetishism in South America (2)*. Chapel Hill, US: University of North Carolina Press.

Weitzner, V. (2012). *Rendición de cuentas de las compañías extractivas en Colombia: una evaluación de los instrumentos de responsabilidad social empresarial (RSE) a la luz de los derechos de los Indígenas y los Afrodescendientes*, trans. M. C. Benítez (p. 179). Bogotá, Colombia: The North-South Institute, Proceso de Comunidades Negras, Resguardo Indígena Cañamomo Lomaprieta.

5 Responses to Mining at a Nasa Indigenous Sub-Regional Organisation

Introduction

Nasa responses to mining are part of the evolving tapestry of Indigenous Latin American resistance practices that have been documented elsewhere (see, for example, Jackson & Warren, 2005; Kearney & Varese, 2008; Bebbington, 2013). I shall illustrate that Nasa responses to mining constitute an institutionalised, everyday resistance, where few acts are construed as mere mobilisation or confrontation. These responses are multifaceted and multilayered. They include a legal dimension, an institutional design, specific projects, territorial control and enforcement, as well as research, dialogue and collaborative advocacy. The Nasa have organised themselves in highly localised Authorities and in a zone Association. As a result, each of the response components can take different configurations when organisational or territorial domains shift. Here, I examine the Nasa responses to mining based on my observations at the Association of Indigenous Authorities of North Cauca (ACIN). In the subsequent chapter, I analyse Nasa responses to mining based on what I learnt at the Canoas Resguardo.

I explain that the Nasa engage in significant, extensive efforts to respond to mining that place strong demands on communities and leaders and that require technical and financial support. I shall concentrate primarily on the legal framework, aspirations and enforcement work because the discourse and intentions within Nasa legislation have not been easy to implement. An overview of the Nasa legal framework illustrates how its diverse interpretations suggest a contested and sometimes deliberately ambiguous position on mining. Various Nasa participation spaces allow the design and debate of these laws. The Nasa economic-environmental system – a system that in Western terms would cover matters such as natural resource management, environmental conservation, sustainable agricultural methods and territorial organisation – is well-established and counts numerous achievements, although it requires technical and legal support. Knowledge, education, dialogue and advocacy activities related to mining are numerous but have not always contributed to shaping Nasa legislation on, and responses to, mining. The arduous work of the Indigenous Guardia, a network of volunteers in charge of protecting territorial integrity, life and

DOI: 10.4324/9781003226895-5

collective and human rights, serves as an enforcement mechanism but could not cope with the extent of unauthorised mining activity. In all, these Nasa responses are significant to the study of Latin American Indigenous struggles with resource extraction and articulates elements that resonate with emblematic cases of Indigenous resistance to extraction.

Two points of clarification about terminology are necessary to reflect the crucial role of spirituality in Nasa understandings of environmental and economic questions, and in guiding Nasa responses to mining. I refer to an economic-environmental system or network and to 'life spaces', following the Nasa terminology that encompasses, but covers much more than natural resource management, environmental and livelihood matters. The Nasa understand economic and environmental processes as one. The banner 'economic-environmental' encompasses work to protect Mother Earth and to promote an Indigenous economy that supports food autonomy. What the Nasa call economic-environmental translates in Western terms to themes such as water conservation, water quality, land management, expansion of the 'Indigenous estate', sustainable agriculture and traditional production methods. When referring to land or water, the Nasa use the term life spaces, which can include them as natural resources but also as spaces within the complex web of life. The name life spaces reflects the Nasa understanding of the spiritual and interconnected dimension of human-nature relationships. The emphasis is people's interconnected and integrated relationship with the life spaces rather than on 'managing' life spaces as external.

Nasa organisations have a wide range of responses to mining that face significant barriers. Over the last decade, national, regional and sub-national Indigenous legislation has opposed mining and provided the basis for activities to control it (CRIC, 2013; ONIC, 2013). However, there is mining in Nasa territories and, as a result, there are tensions and opacities. The material circumstances of some of the communities, and a political-economic environment that is not supportive of Nasa views on the economy and the environment, hinder the efforts of Nasa organisations in implementing the NO mandate. Rather than an exhaustive inventory, what follows is a snapshot of prominent elements of Nasa responses to mining during the transition from Peace Agreement negotiation to implementation. The discourse and intentions within Nasa legislation have not been easy to implement, so this discussion mostly describes the legal framework, aspirations and enforcement work.

I begin by summarising the Nasa legal framework and how its diverse readings demonstrate a contested and sometimes deliberately ambiguous position on mining. Following this, I describe the Nasa economic-environmental system and its funding, technical and legal support needs. Later, I cover some of the knowledge, education, dialogue and advocacy activities related to mining that I learned about. I interrogate the role of these activities in shaping Nasa legislation on, and responses to, mining. I next reflect on the work of the Indigenous Guardia and illustrate the tensions and collisions the

Guardia witness as they move through the territory implementing control responses to mining-related problems. I shall finalise by discussing how the Nasa resistance builds and expands on common elements of the Latin American Indigenous people's resistance to mining, while displaying great innovation thanks to a favourable legal framework and a distinct Nasa organisational process.

Indigenous Law, Legal Action and Advocacy

Colombian Indigenous communities have legal autonomy and issue Indigenous legal mandates. Indigenous mandates are legal decisions with a territorial scope that aligns with that of the issuing Authorities, i.e. local (Cabildos), zone (zone associations like ACIN) or region (organisations like CRIC). Every four years CRIC, and its affiliated zone associations such as ACIN, hold Congresses, assemblies of as many as 10–15,000 people, where the Authorities and communities issue mandates marking the pathway for the subsequent four years. After several months of preparation, the Congresses implement iterative, participatory processes where technical teams, communities and Authorities debate and gradually refine proposals for Indigenous mandates. Some mandates deal with territorial dynamics that, like mining, affect life spaces. In the last 20 years, the mandates of Cauca Indigenous Congresses have leaned towards a position of NO to mining in Indigenous territories, in particular if it is practiced by outsiders, and have called for clearer and more stringent regulation of mining from Cabildos (CRIC, 2013). These mandates state the predominant political and moral position. However, some Nasa or their Authorities participate in mining. This reflects different interpretations and contestations of the mandates.

The moral-economic tension about mining is a continuum that runs through most aspects of Nasa responses to mining, beginning with interpretations of Indigenous law. I heard many different readings of the mandates on mining, ahead of the 2017 Congresses. For example, NasaUs, an ACIN Councillor, said the position on mining is a clear NO:

> *We had the Congress of Jambaló in 2002 and another in Coconuco in 2011. We gather in large numbers at those Congresses. We put mining on the table, and the community mandate was that mining in Indigenous territories is a no go. That's because we have another concept, the minerals are like the energy that is part of our lives. Seen from outside the Indigenous community, people might call it the patrimony of the communities. So we say there will be no exploitation, not artisanal, nor with machinery. It's a no go. It is not a matter of structure, ACIN cannot say mining can be done here or there, CRIC cannot say mining can be done here or there. It is a mandate. Mining will not be done. That is what we fight for.*

Another senior ACIN leader, El Cacique (pseudonym), thought the position was more nuanced because there were Indigenous Authorities doing mining and that presented a contradiction:

> *Various regional and national Indigenous platforms have issued mandates on mining. One is the NO to mining. There are zone level mandates to defend the territory and life spaces. So within that is the mining issue, but it is not explicit there. The regional and the national mandates are both a NO to mining. But mining is practiced in the territories with State recognition: limestone in Tacueyó, and gold in Delicias. Then we use other minerals. For example, ceramics. So we need to look at other minerals beyond gold. In the Congress, we will look at this problem and see what the mandate is. We'll see if we can frame the territorial control exercise within our law. Right now, we cannot say no to mining, when we know that not only gold but other minerals are being mined. We need clarity and we will take the problem there and let the Congress make a decision.*

There was tension between the political and moral positioning that sees any mining as morally wrong, and an argument grounded on Nasa economic activity. Mandates emanating from the 2017 Congresses of ACIN and CRIC that happened after I concluded the fieldwork might have addressed some of these tensions. The 2017 Mandates encouraged the control of mining in order to "*liberate*" and protect Mother Earth and the communities (ACIN, 2017; CRIC, 2017). However, it was unclear whether the decisions would be fully implemented because ACIN's 2017 Congress had low attendance, and participation is the basis for legitimacy. Let us briefly review the results of the 2017 Congresses.

The 2017 ACIN Congress of Tóez

ACIN's third Congress, held in 2017 in the Indigenous Resguardo of Tóez (Caloto, Cauca), identified mining as an "*extermination strategy*" against communities and their territory that, when combined with crops of illicit use, the armed conflict, modernity and family disharmony, were changing family structures that are the fundamental basis of the Indigenous organisations (ACIN, 2017). The Tóez Mandate confirmed the NO to mining, citing concerns about "*the inappropriate use of the territory, damage to sacred sites and water springs, and the use of mercury that causes illness and imbalance in all beings that are part of nature*"; and ongoing threats against those who opposed mining (ACIN, 2017). It issued decisions to block attempts at prior consultations and to seek reversion of mining titles granted without consultation; and called for ongoing work to create sustainable and clean agricultural and pastoral production initiatives as alternatives to mining and crops of illicit use. The Congress also provided guidance for Indigenous Authorities to exercise environmental functions and control mining through institutions, legislation and projects (ACIN, 2017). The Congress called on Authorities to understand territorial organisation not only in terms of economic activity but considering "*how Mother Earth is organised, the life spaces and the spiritual order*". It stated the response to mining must be based on

Nasa ethical and spiritual principles, but also offer concrete solutions (ACIN, 2017). It was overall a call to action and to increased environmental responsibility from Cabildos.

The 2017 CRIC Congress of Sotará

The Mandate from CRIC's 2017 Congress, held in Sotará, classed mining in Indigenous territories as a moral failing and a contravention of Indigenous mandates that weakened Indigenous autonomy and capacity for self-government (CRIC, 2017). The Mandate re-stated previous decisions to reverse unconsulted mining titles and block consultations, and called for increased and collaborative territorial and political responses to mining (CRIC, 2017). The Mandate encouraged legislation on the status and rights of territories and life spaces, such as water and other natural resources, linking them to Indigenous collective rights and law. This resonates with legal developments in Colombia, through the Constitutional Court, as well as Ecuador and Bolivia (Sieder, 2007; Velasco, 2011). It also reflects environmental concerns about a range of economic practices besides mining such as agriculture and tourism. The Mandate declared "*sacred territories and special life spaces free of impacts, use and exploitation that might contravene the permanence and survival and the rights of Mother Earth*" (CRIC, 2017). The Mandate declared water as a "*living and spiritual being, a rights subject within the framework of collective rights of Indigenous Peoples [...] not to be subjected to commercialisation or any use different to community use*" (CRIC, 2017).

The mandates are central to the construction of Indigenous identity and to the tradition of Indigenous legal advocacy. Throughout its text, the Sotará Mandate illustrates that concerns about economic activity originate in its links to territorial integrity and Indigenous identity. For example, when saying:

> *Mother Earth has an order. We must achieve equilibrium to live in harmony. We must administrate from a legitimate collective perspective, through Life Plans, to continue along the path we have mapped out, based on the principles and platform of CRIC, we must strengthen our identity, and the good use of the Indigenous economy because economic resources are a means to strengthen ourselves.*
>
> (CRIC, 2017)

These mandates illustrate how responsible economic practice is part of Indigenous identity. Mining as an economic, environmental, spiritual and territorial phenomenon is a source of tension driving debate and reflection, and testing the principles behind Indigenous identity. The Indigenous Congresses are spaces of collective resistance, where communities recreate and reaffirm their identity, but also contest it. Nasa collective action provides a space for identity generation, affirmation and transformation (González Rosas, 2016). Indigenous Congress mandates, together with national and international treaties and legislation, also provide a platform for legal responses to mining, in line with the Nasa legalistic resistance tradition (as described by Rappaport, 2000).

Institutional Designs

Formalising the environmental functions of Cabildos has been a CRIC proposal since its 2005 Congress, where the 122 affiliated Indigenous Authorities ratified the role of Indigenous Authorities as Territorial, Economic and Environmental Authorities. Here I examine this proposal, with the caveat that, while this had been mandated, the shape of its implementation had not been consolidated. There were a range of evolving views internally and contestation from the National Government.

Several participants argued that Indigenous Authorities already performed environmental functions but lacked State recognition, funding and support. For example, an environmental expert from ACIN stated:

> *We have always been Territorial, Economic, Environmental Authorities, but the State has not recognised us. In this struggle, in our ongoing march, we are seeking recognition from the State. For millennia, the Indigenous people have preserved the territory. We have defended it. The environmental authorities haven't given us any help, even though they have salaries. They have never lent a helping hand. They have never come forward to support the Authorities and the communities. It is the communities that have put their lives on the line to defend the resources, or the water sources [...] what we call life spaces.*

Nevertheless, there are obstacles to exercising those environmental functions. A national legislative process has been underway for some years to regulate the role of Cabildos as Environmental Authorities. However, conversations with members of ACIN indicated that during 2017 Government drafts had watered down the proposed territorial autonomies. This and other disagreements with the National Government triggered large-scale Indigenous movement protests during 2017.

The proposal aimed at:

- Supporting the application of Indigenous justice by Cabildos
- Improving knowledge and implementation of Indigenous law, and continuing the 'Liberation of Mother Earth' process
- Reducing external market reliance; protecting water, traditional foods and seeds; and promoting clean and diverse agricultural production within families
- Promoting harmonious relationships with the territory, supporting collective approaches, solidarity and the fair distribution of lands and benefits.

(CRIC, 2016)

A participant from CRIC explained to me that the following were necessary to consolidate the proposal: i) a normative framework; ii) work to strengthen the Indigenous economic-environmental system; and iii) guidelines, mandates and activities to protect Indigenous territories and life spaces. In the following sections, I discuss some of the complexities of the normative framework, as well as the economic-environmental system. The work to protect life spaces and territories is holistic and I cover it throughout this chapter.

Normative Framework: Opportunities and Uncertainties

The legal framework relevant to the proposal of Cabildos as Territorial Economic Environmental Authorities is a foremost concern for the Nasa. Legalism has been a central element in the Indigenous struggle for autonomy (González Rosas, 2016; Rappaport, 2000). The legal framework defines the possibilities of the proposal. The enabling framework for the proposal includes international treaties such as the UNDRIP (2007), Colombia's National Constitution (Political Constitution of Colombia, 1991) and laws, as well as Indigenous mandates. In parallel, according to CRIC's analysis, some national laws have given and taken away autonomies, or left questions unresolved. This led the Indigenous movement to put forward a legal proposal to clarify the environmental responsibilities of Cabildos. This was at the centre of tensions with the National Government at the time of this study.

The legal matters that the Indigenous movement wanted to see resolved were many. Indigenous analysts cited delineation of forestry reserves, parks and *páramo* (sub-alpine environments, rich in water sources) overlapping Resguardos; agrarian reform that overlaps Indigenous autonomies (see Law 200 (1936), and Law 100 (1944)); as well as uncertainties about the role of Indigenous Authorities in the National Environmental System (see Law 99 (1993)). Two national legislative developments had increased interest in the environmental role of Cabildos among the Indigenous participants of the study. Decree 953 (2013) regulated payments for environmental services in areas surrounding municipal water supply sources, opening opportunities and raising questions about the role of, and payments owed to, cabildos as guardians of municipal water sources. Historically, Indigenous peoples in Cauca have been pushed away from the flat and fertile lands of the inter-Andean valleys (Rappaport, 2000; González Rosas, 2016). They were driven to the mountaintops that are home to municipal aqueduct water intakes. For example, the Canoas Resguardo harbours the water intake for several local aqueducts. Canoas acquired surrounding land and dedicated it to conservation. Both ACIN and CRIC planned to characterise such services and explore the implications. A second significant legal development for the participants, Decree 1953 (2014), called Cabildos to define, execute and evaluate environmental policies "*within the framework of their Life Plans in the relevant territories, national legislation and consistent with principles of coordination, concurrence and subsidiarity*". There was room for interpretation on what those principles meant in the context of centre-region-locality tensions on territorial organisation, in particular with natural resource/life space concessions at stake. All these matters were to remain subject to interpretation for as long as the environmental functions of Cabildos were not clarified.

The Indigenous Economic-Environmental System

Nasa organisations have the institutional and policy architecture for an Indigenous Economic-Environmental System but lack funding and a supportive Government policy environment to implement it. What mainstream Colombian society calls the 'economic model' the Nasa refer to as the 'economic-environmental system'. Central to the Indigenous economic-environmental system is family-based, biologically diverse and traditional agricultural production. Timaná, an environmental specialist, explained that once traditional Indigenous agriculture was dismissed as outdated, but now science favours it because of its positive interactions with biodiversity, soil equilibrium and the water cycle, which contrast with the deleterious effects of intensive livestock and monoculture in Cauca. During the 1970s and 1980s, substantial community and international resources went into the design of sustainable agricultural methods based on traditional knowledge. One of the results was the *Tulpa*, a one-hectare parcel with over a thousand species that mimics naturally occurring biodiversity while producing a balanced diet for an Indigenous family. The Tulpa is the form of production that Indigenous Authorities promote. The Indigenous Economic-Environmental System supports traditional agricultural production methods, community production projects, value adding activities, logistics and commercialisation. In contrast, the predominant policy environment favours large-scale agriculture and a monoculture approach to rural development (see Government of Colombia, 2013).

Activities to strengthen the Indigenous economic-environmental system can result in high quality products that reach beyond the local Nasa markets. For example, at the time of the study the Northern Zone had a trout farm, a dairy products enterprise, a fruit juice factory and a traditional medicine factory; and ACIN ran a commercialisation centre in Santander de Quilichao. However, sometimes production was not sufficient to meet Northern Zone demand and the unreliable energy supply caused difficulties with refrigeration and production losses. Nevertheless, some shelf-stable traditionally grown Nasa products were reaching larger markets and were recognised for their quality in Bogota and in the coffee circles.

The expectation was that Indigenous agriculture projects would provide alternatives for Indigenous miners. However, there were difficulties. El Cacique warned that there were no political, judiciary or financial guarantees for livelihood substitution projects for Indigenous miners. He believed there always had been other livelihood options, but the lack of support made them difficult to implement: "*It is very difficult when the National Government commits to a national policy of mining development. People will even kill each other to do mining, because it gives them a profit*". Analyses by ACIN and the

Cauca Regional Attorney's Office concluded that agriculture and livestock earnings are minimal compared to those of coca, marijuana or mining. In addition, El Cacique was concerned about the direction National Government policy had taken when it came to rural development. In his opinion, the rural development plans and mining policy were desktop-based and in tension with Indigenous principles. Frustrated, El Cacique said:

> *For example, we don't have funds to help someone who wants to grow blackberries, we have no money, or for livestock, we have no money. The Government won't help us. We could do a harmonisation, education, spiritual, cultural and autonomy process, but there will always be the demand from people and where do I get the funds to fulfil their demands. That is the big challenge for ACIN and the Authorities. We have a big dream, big expectations, but we won't be able to make it a reality, we won't be able to give quick responses to community demands.*

El Cacique's analysis highlighted significant barriers. Ultimately, behind the limited resources were the socio-economic disadvantages of the communities and the fact that these communities and the Government had different ideas about rural development.

Knowledge, Dialogue and Education

ACIN participants spoke to me about knowledge, dialogue and education activities that contributed to debates and awareness about mining. These happened either as collaborations with external parties or as internal activities within ACIN or the Resguardos. The activities included record keeping and documentation, dialogue about mining, advocacy efforts, and research or analysis projects. These activities seemed somewhat disconnected, with the links built on an as-needed (urgency) basis, rather than a structured organisational response to mining. However, they had contributed to the education of local Indigenous mining experts and to raising awareness about mining internally and externally.

ACIN and the Canoas Resguardo had conducted work to better understand mining dynamics. The process had been organic with some attempts to scale it into an institutional response. Several research activities had provided information and skills to local researchers and access to external expertise and collaborations, besides producing reports and cartographic analyses. However, the research projects had not always built on each other nor had the results been analysed as a whole. Neither did I find evidence of them informing other responses to mining. Most mining research was done from *Casa de Pensamiento*, ACIN's research arm that, according to several participants, was closed because of internal political and gender dynamics. So, the research it supported came to a halt and it was not used to inform ACIN mining work or shared with Resguardos. As a result, research continuity sprung from the commitment of individual researchers, internal or external. Whether new research was initiated was highly contingent on external interest, the ability to access funding, local and organisational political will, and the incidence of mining conflicts. In the absence of a sustained approach to mining research, it had been difficult for research activities to contribute to transforming mining conflicts. Lack of clarity on research priorities, in a context of escalating mining conflicts and changing mining dynamics, made it difficult for research to: i) provide timely information to prevent conflicts, ii) access field data and iii) respond to immediate conflicts as well as longer-term questions.

Research participants spoke at various points about dialogue activities aimed at deterring Nasa participation in mining that had not yet achieved the desired results. However, I did not observe these during the fieldwork. ACIN and some Resguardo leaders saw the dialogue activities in a different way from the miners I interviewed. A senior leader at ACIN told me about some of the dialogue ACIN implemented when external miners attempted to recruit Nasa people in order to bypass the Nasa Authorities. He explained:

> *We begin to raise awareness about the effects on the waterholes, on the daily lives of our families, on health matters. People often say that there will be a management plan for the mine, but we know that even a great management plan can only mitigate rather than eliminate the impact. We know there will be tailings dams and that chemicals like cyanide accumulate and affect our genetics. We discuss all of this with our communities. We emphasise that mining can affect the water sources, and there are places that cannot be touched without certain conditions being met, so if mining is done, it will affect the waterholes. We get our masters of traditional knowledge to explain this situation to people.*

I heard similar narratives from other participants from ACIN and from Resguardos – that dialogue was the first response to mining by Nasa people. However, what dialogue entailed was contested. The miners I interviewed did talk about discussions where Authorities spoke about principle-based positions and the deleterious effects of mining, but the miners believed the negative side was overplayed and felt that economic alternatives were not available. It appears that for dialogue to help the communities in resolving the grievances and disagreements surrounding mining, it would be necessary to have data on the effects of mining as well as alternatives to the alleged lack of livelihoods.

Education initiatives had not always been successful in deterring young people from working in mining. At the time of the fieldwork, ACIN was implementing an education and social mapping initiative centred on water. Water is a less confrontational theme, but also an ideal proxy for mining. According to Yu (pseudonym), ACIN researchers invited four Resguardos with mining or crops of illicit use to participate. The Canoas and Huellas

Resguardos came on board while Delicias and Munchique, both with a substantial mining presence, did not. The program concentrated on the cultural and political understandings of water as a life space. Yu said it was attracting interest amongst young Nasa, but was concerned that it could eventually mirror the experience of other education programs where "*people would finish their training session, and then return to their mining work*". However, without data on mining activity, it would be difficult for ACIN to assess how effective their initiatives are.

In summary, there had been several efforts by ACIN and its affiliated territories to understand and debate the workings of mining, but these activities were yet to be connected into a coherent whole. There were difficulties in ensuring support for research in a highly politicised Indigenous organisation. There were barriers to education and dialogue in Resguardos where mining and associated conflicts were present. Added to resource scarcity and to the technical, moral, social, safety and emotional difficulties and tensions that an Indigenous researcher needs to balance to look at mining in depth, these hindered the possibilities for research to fully serve the communities and Authorities. The organisational dynamics decreased the chances of continuity and political support for research. Without a clear champion and resources for mining-related research, dialogue and advocacy, ACIN affiliated Resguardos would struggle to build on one another's experience and gather sufficient information to debate with the proponents of mining internally and externally. Nevertheless, it is remarkable that research was being conducted, mostly thanks to the support of external donors and individuals within ACIN and the Resguardos. Dialogue, on the other hand, seemed to take place as a corrective measure, more than through proactive initiative. This was an area where some miners voiced concern and frustration and where more effort was necessary.

Territorial Control: The Work of the Guardia

> *The Indigenous Guard is aware of the threats against Authorities, of the people from elsewhere that are encouraging mining. The miners are not just Indigenous people. There are other people promoting capitalism. They are elsewhere. We look at that. We look at the territory, make a presence there and report back on what is happening. The Indigenous movement denounces illegal mining and the Government says, 'yes, we know, we are investigating it'.*
>
> (NasaUs)

The mandates, education and alternative livelihood projects were not sufficient to respond to mining given the socio-economic context, so a lot of enforcement work was necessary. This territorial control and defence work is in the hands of the Indigenous Guardia. Charged with those responsibilities, the Guardia face the community divisions, tensions, struggles and unity around mining questions, directly and in their most concrete expressions. The

Guardia is a volunteer network implementing large-scale activities at significant cost to its members and their Nasa organisations. It faces great barriers, tensions and dangers.

The Guardia works to control mining within the scope of Indigenous mandates and their various local interpretations. NasaUs (pseudonym), an ACIN Councillor, explained to me that the Guardia must enforce a ban on mining. Concerned about mining being out of control, the Councillor told me:

> *We are living this situation where families are doing this work and, let's say, they don't want to listen. They are bringing difficult situations to the territory. For example threats, even deaths. Yes, they have caused deaths! That is why we say that mining is forbidden in the territory. That is the mandate and even if people want to do it, the mandate is what emanated from the Congresses. So we must do the work.*

On the other hand, Carlos (pseudonym), a Guardia leader, was careful to differentiate between what he saw as mining activities that Indigenous Authorities allow and are not deleterious, and those involving heavy machinery owned by powerful external forces. Having inspected the sites in person, Carlos could readily describe the differences:

> *Suppose it's artisanal mining. No one will oppose that. Working with a pan and a shovel, the miner works to find the gold, he may or may not get it. But a digger, or two, or three, that is really complicated. Wherever you see one digger, you find one to two hundred people panning for gold.*
>
> *Is an informal miner going to work with machinery, four or five diggers? Sometimes they say 'we are starving'. So we ask, how can they get five diggers? How much is a digger? No one with limited resources is going to venture out with five diggers into a territory where they can get confiscated. They know we can confiscate them!*

Carlos empathised with the fellow community member seeking an independent livelihood, but had no time for others he saw as pawns of exploitative forces. When machinery is involved, the Guardia destroys[1] or confiscates it.

Differences of interpretation aside, the Guardia had to patrol the territory, respond to community concerns and act to stop unauthorised mining. This included deterring "*multi-nationals*" and external small-scale miners to prevent them from mining and to deter Indigenous recruitment. Carlos had observed that the difficult material circumstances of the communities allowed external mining entrepreneurs to consolidate the recruitment of locals as a preferred 'divide and conquer' strategy. Carlos told me:

> *A strategy they [mining entrepreneurs or companies] are using is to get small scale miners to enter the territories, hopefully they don't get stopped,*

> *and the important thing is they will gain supporters. They gather support, and then when the Cabildo says something, they reply: 'have a look, the local people work here, so how are you going to try to stop this? What jobs are you going to offer them?'*

Apparently, this divisive strategy had worked. Many participants referred regretfully to an area called El Nilo, in the municipality of Caloto, where an Indigenous person dared the Guardia to burn an illegal mining digger with him in it, and where there have been confrontations between the Indigenous people and the Guardia.

As mining companies and entrepreneurs of various scales sought entry into Nasa territory, the Guardia acted to block attempts at consultations, in line with Indigenous mandates. The Guardia's control served to pre-empt what Carlos stated was a common practice of claiming that an FPIC process took place when there was no more than a conversation with community or Cabildo members. What this meant in practice is that the Guardia denied entry or escorted out mining company personnel and mining entrepreneurs. Carlos explained that:

> *We've had a lot of difficulties with the multinationals, because they always start talking about prior consultation. [...] They come and start talking like you and I are doing, but instead of having the conversation, 'What do you think? This is what I think and so on' they say 'We'll come. We'll build you a road. We can fix your school. We'll give you a big morning tea. Sign here.' That's prior consultation for them. We always said 'you must consult us about the effects of the proposed mine!' So when they send their personnel, we have to stop them and say: wait a minute, no more 'prior consultations' here. We've had to escort people out of the territory.*

Having witnessed some of the Guardia's work, such as in the working bee I wrote about at the beginning, and just by counting *chivas* (traditional buses) and people leaving the ACIN headquarters for mingas (working bees) against mining, I became curious about the scope of the Guardia's efforts. I asked Councillor NasaUs, who estimated that, in 2016 alone, the Nasa gathered in seven mingas to control mining, each with at least 1,000 participants and lasting between three and eight days. The Nasa territories are spread throughout a large area spanning nine municipalities. People came from all over and needed transport, they walked long distances, they camped and had to be fed. A minga of such scale involved significant planning and logistical effort, and the costs were far from negligible, given the material conditions of the Nasa communities. Considering only the time communities put into a minga, assuming short days of eight hours, well below Nasa practice, the communities were investing between 24,000 and 64,000 hours in each minga. Even at the low wages that local agricultural workers earned in the region at the time, the sheer number of hours has a significant opportunity cost. At

$25,000 COP daily wages, the foregone income per minga could amount to between $75,000,000 and $200,000,000 COP, or $25,000 to $66,000 USD based on 2016 costs and exchange rates. This is a total of $175,000 to $463,000 USD in 2016 alone. A cursory and extremely conservative estimate of direct costs such as transport ($20,000 COP) per person and food ($20,000 COP a day) would more than double the costs.

This large-scale and costly work presented the Guardia and communities with numerous difficulties. Given the size of Nasa territories, their mineral endowment, the material poverty of the local communities and high levels of illegal and criminal armed group presence, the permanent Guardia core was insufficient to patrol the entire territory. Furthermore, the Government was unable to deter the advance of illegal miners. The Nasa hoped that the National Government would re-allocate funds, saved from declining military investment in the wake of the Peace Agreement, to support the work of the Guardia. The political environment in which this study took place made that prospect unlikely. However, at the time of writing, the political landscape had changed drastically. Regardless, physically stopping miners does not permanently deter them. Carlos believed that the territorial control needed to be complemented with more education and livelihood projects, because the work of the Guardia, alone, was not enough. At times, technical issues prevented the Guardia and the Indigenous Authorities from enforcing Indigenous mandates. For example, earlier on, I spoke about the permits, logistics and skills required to block mining tunnels during a minga. These were not readily available within the community, so action was delayed, causing significant frustration, not least because of the large opportunity costs (foregone income) of participating in a minga.

Assuming that mining was temporarily controlled in one Indigenous territory, neighbouring Authorities could still be persuaded to allow mining or fail to halt it, rendering the environmental controls of one territory ineffective, as pollutants and social impacts know no municipal or territorial boundaries. When Authorities were approached one by one, their ability to resist the illegal mining wave could be overcome by financial needs and fear. Gradually, illegal miners acquired more supporters and controlled larger areas. Just like "*the Guardia is the community*", the miners could become the community and intimidate Authorities. These dynamics had put the Guardia in situations where the miners claimed the Guardia was an obstacle to industry and livelihoods, or simply that an Afro-descendant, Indigenous or Government Authority provided permission, causing delays to Guardia enforcement regardless of the veracity of the claims. Finally, when Authorities decided to withdraw their support for miners, it was difficult for them to confront their own community and the powerful agents that controlled illegal mining.

External parties offered valuable support to the Guardia work, but in some cases this caused tensions about Indigenous autonomy. This is not exclusive to the Nasa, but an element of the Latin American Indigenous peoples' work to decolonise cultural and political relationships which requires "*the negotiation of autonomy with national and international entities*" (Kearney and Varese, 2008, p. 202). The UN human rights delegates would at times accompany Guardia actions as observers. Having identified mining as a key threat during the post-conflict transition in Cauca Indigenous territories, the UN was supporting community work to assert community constitutional autonomies and denounce environmental and human rights incidents. The Regional Attorney's office had begun collaborating with the Nasa Authorities and Guardia on law enforcement and investigation including illegal mining. However, this particular collaboration raised questions about Indigenous autonomy because it involved an external intervention. In the absence of the technical skills and necessary permits, alternative scenarios are difficult to envisage. In all, perhaps autonomy is not an antonym to collaboration; perhaps stronger levels of autonomy emerge from utilising the best available resources and expertise.

The Guardia's work was one of the most arduous and consuming efforts to respond to mining that I became aware of. It was a central element in Nasa responses to mining and exemplified the tensions, difficulties, dangers and innovations involved in responding to mining in post-conflict North Cauca. The Guardia mobilised, on an ongoing basis, significant in-kind community resources, performing a mammoth yet nimble effort in an environment of material poverty and State absence. However, their effort alone could not halt unauthorised mining. This effort needed to be complemented with external resources, skills, livelihood projects, education, spiritual work and collaborations. In their territorial journeys and as they met the changing faces of mining, the Guardia experience revealed the tangible tensions inherent in territorial autonomy in a resource rich and conflict-affected region. These experiences suggest that only responses as complex as the mining system itself will provide answers to the problems and tensions facing the Nasa.

Regional Commonalities and Nasa Innovations in Resistance

The significance of the Nasa response to mining to the study of Indigenous struggles with extraction in Latin America is manifold. Latin American Indigenous resistance frequently includes large-scale mobilisations and legal action (Bebbington & Scurrah, 2013, Jackson & Warren, 2005; Kearney & Varese, 2008). Where extraction is concerned, emblematic cases of resistance, for example in the Peruvian and Ecuadorian Amazon as well as in Guatemala, have involved not only large-scale protests, but asset destruction or operational disruption, complemented with international legal cases, to exert pressures on national governments and extractive companies (Arellano-Yanguas, 2013; Jackson & Warren, 2005; Kearney & Varese, 2008; Sieder, 2007).

The Nasa response to mining builds on these 'regional' elements of resistance, while also exemplifying the possibilities of institutionalised resistance in a favourable legal environment. The Nasa can issue autonomous law. The Colombian legal framework allows their judicial autonomy to prevail "*to the greatest extent possible in Latin America*" (Stavenhagen, 2002 in Sieder, 2007).

Consequently, when mining actors breach Indigenous law, they can be sanctioned. The Nasa have an established implementation and enforcement capacity through ACIN. An emblematic example of Indigenous resistance to extraction, the Cofan of the Ecuadorian Amazon, channelled their resistance to oil extraction through an environmental NGO because their Indigenous organisation was for many years unable to gain legal recognition domestically, and relied heavily on the profile of an unusual leader (Cepek, 2012). In contrast, the Nasa have a 24-year-old organisation with legal personhood, recognised organisational capability, and an established workforce and network of practitioners. The Nasa resistance to mining also takes a particular dimension in that it responds to illegal actors and to miners' recruitment of Indigenous people, where territorial control and enforcement mechanisms like the Guardia become essential and require a day-to-day presence for monitoring and deterrence, in addition to large-scale actions.

There are other parallels and similarities between the Nasa resistance and those of other Indigenous peoples in Peru, Ecuador and Bolivia. In terms of similarities, collaborating with local and international movements and environmental organisations has been a common element of Indigenous resistance to extraction in Peru and Ecuador as well (Bebbington, Humphreys Bebbington, Bury, Lingan, Muñoz, & Scurrah, 2008; Pratt, 2013; Bebbington, 2013; Bebbington, Scurrah, & Bielich, 2011; Rival, 2013), but it is not always the case that Indigenous organisations have the institutional capacity to make international collaborations effective (see Pratt, 2013). Like the Nasa, Indigenous peoples in the Peruvian Amazon have followed a gradual process of documenting complaints and seeking dialogue, ultimately resorting to mobilisations and blockades to bring an unresponsive State to the table (Bebbington, Scurrah, & Bielich, 2011; Bebbington & Scurrah, 2013; Rival, 2013). Other Indigenous resistance processes have also faced divisions, where sections of the Indigenous population participate in mining and are more amenable to the extractive projects of the State (see, for example, Warnaars & Bebbington, 2014, on the case of El Pangui). Conflicts around extractives involving Indigenous people have led some institutional innovations on the management of extractive projects or of the conflicts themselves in Peru, Ecuador and Bolivia (Bebbington 2013, Crabtree & Crabtree Condor, 2013; Rival, 2013). However, in the specific case of the Nasa resistance to mining, institutional innovations outside the Indigenous movement are less discernible. Perhaps one of the most common elements to these resistances against extraction (or foreign-led extraction) is the blunt, violent repression they have faced (see Arellano-Yanguas, 2013; Perreault, 2006), which despite losing effectiveness in some countries from the perspective of the perpetrators (Crabtree & Crabtree Condor, 2013), continued to advance and undergo mutations in post-agreement Colombia.

The Nasa have devised a wide scope of responses to mining that either target or circumvent tensions about mining within and without the Indigenous movement. First is the tension between the political and moral positioning of the Nasa, and the pressures resulting from historic dispossession and inequality. Second, the problems that result from absence of: a) rule of law to control mining; b) guarantees for Indigenous FPIC rights; and c) coordination between overlapping or neighbouring local, regional and natural resource management authorities. Third, autonomy tensions when accessing external funding and expertise to respond to the difficult technical questions that a complex economic system like mining presents.

These tensions permeate Nasa responses to mining, affecting their effectiveness, credibility and, ultimately, resonance with the Indigenous moral values. If left unresolved, the tension between morality and political positioning on the one hand, and livelihood dilemmas on the other, has the potential to fragment Indigenous communities. A nuanced response to moral-political-economic tensions is contingent on reliable information about mining, based on traditional and Western knowledge and expertise, that can enrich dialogue and decision-making within communities and with external decision-makers. For this, it will be necessary for the Nasa to define the role that intercultural, interdisciplinary research has in their responses to mining.

The Nasa have extensive experience in Participatory Action Research, based on the methodology developed by Orlando Fals Borda (1979, 2013) in collaboration with grass-roots members of Colombian rural communities. Participatory Action Research sees academic and grass-roots researchers collaborate in analysing questions of interest to communities, and design actions that raise awareness and affect change (Rappaport, 2020, Pereira & Rappaport, 2022). The Casa de Pensamiento, ACIN's former research unit which is no longer active, conducted several research exercises on mining with some results published. For example, a study by Caro Galvis and Valencia (2012) appeared in the Señas Magazine once published by the Casa de Pensamiento. These exercises demonstrate the research efforts. What was unclear was how ACIN was utilising research findings.

Research matters aside, there were encouraging signs that Indigenous and State authorities were beginning to respond collaboratively to illegal mining at the time of the study. However, questions of funding and access to technical expertise for Indigenous communities remained. Ideologically, the National Government of the time and Nasa Authorities inhabited diametrically opposed spaces. This raised questions about the role of State actors in mining governance in Indigenous territories and about how authorities collaboratively governed mining while safeguarding Indigenous autonomies. At the time of the research, legislation on the environmental competencies of Cabildos had not been passed, which signalled a long negotiation ahead.

The post-conflict transition, including FARC's return to civilian and political life and the increasing levels of armed criminal activity, will make these questions and tensions more complex and responses harder to implement. In this rapidly changing context, how do these Nasa responses to mining articulate with peacebuilding? The next chapter will provide some examples of the lived tensions in the Canoas Resguardo. It will complete the foregrounding necessary for

Chapter 7, where I shall discuss how responses to mining intersect with peace-building in a conflict-to-peace-transition context.

Note

1 This is a law enforcement measure applied also by the police and army, where remoteness prevents machinery confiscation.

References

ACIN. (2017). *Documentos sobre el Mandato del III Congreso Cxab Wala Kiwe-ACIN en Tóez*. Santander de Quilichao, Colombia. ACIN.

Arellano-Yanguas, J. (2013). Minería y conflicto en Perú: Sembrar minerales, cosechar una avalancha de piedras. *Industrias extractivas, conflicto social y dinámicas institucionales en la Región Andina*, 151–184.

Bebbington, A. (2013). *Industrias extractivas, conflicto social y dinámicas institucionales en al Región Andina*. Lima: IEP, Instituto de Estudios Peruanos.

Bebbington, A., Humphreys Bebbington, D., Bury, J., Lingan, J., Muñoz, J. P., & Scurrah, M. (2008). Mining and social movements: Struggles over livelihood and rural territorial development in the Andes. *World Development*, 36(12), 2888–2905. doi:10.1016/j.worlddev.2007.11.016.

Bebbington, A., & Scurrah, M. (2013). Hydrocarbon conflicts and Indigenous peoples in the Peruvian Amazon: Mobilization and negotiation along the Río Corrientes. In A. Bebbington & J. Bury (Eds.), *Subterranean struggles: New dynamics of mining, oil, and gas in Latin America* (pp. 173–196). Austin: University of Texas Press.

Bebbington, A., Scurrah, M., & Bielich, C. (2011). *Los movimientos sociales y la política de la pobreza en el Perú*. Lima: IEP, CEPES, Grupo Propuesta Ciudadana.

Caro Galvis, C., & Valencia, Y. (2012). El caso de pequeñas y medianas minerías en el Cauca: ¿Alternativas o amenazas a la autonomía indígena?. *Señas*, 2, 17–27.

Cepek, M. (2012). *A future for Amazonia Randy Borman and Cofán environmental politics*. Austin: University of Texas Press.

Crabtree, J., & Crabtree Condor, I. (2013). La política de las industrias extractivas en los Andes centrales. In A. Bebbington (Ed.), *Industrias extractivas, conflicto social y dinámicas institucionales en al Región Andina*. Lima: IEP, Instituto de Estudios Peruanos.

CRIC. (2013). *Mandato del XIV congreso del CRIC en Kokonuco*. Retrieved from www.nasaacin.org/index.php/informativo-nasaacin/3-newsflash/5868-mandatos-del-xiv-congreso-regional-del-cric.

CRIC. (2016). *Unidad, Tierra y Cultura*. Autoridad Territorial Económico Ambiental.

CRIC. (2017). *Documentos sobre el Mandato del XV Congreso del CRIC en Sotará*. Santander de Quilichao, Colombia: ACIN.

Decree 953 of 2013, reglaments article 111 of law 99 of 1993 (2013).

Decree 1953 of 7th October 2014 (2014).

Fals Borda, O. (1979). Investigating reality in order to transform it: The Colombian experience. *Dialectical Anthropology*, 33–55.

Fals Borda, O. (2013). Action research in the convergence of disciplines. *International Journal of Action Research*, 9(2), 155. https://doi.org/10.1688/1861-9916_IJAR_2013_02_Fals-Borda.

González Rosas, A. M. (2016). *Vivimos Porque Peleamos – Una mirada desde abajo a la resistencia indígena del Cauca, Colombia.* Ciudad de México: Memorias Subalaternas.

Government of Colombia. (2013). *Plan Nacional de Desarrollo 2014–2018.* Bogotá, Colombia: Government of Colombia.

Jackson, J. E., & Warren, K. B. (2005). Indigenous movements in Latin America, 1992–2004: Controversies, ironies, new directions. *The Annual Review of Anthropology*, 34(1), 549–573. doi:10.1146/annurev.anthro.34.081804.120529.

Kearney, M. G., & Varese, S. (2008). Indigenous peoples: Changing identities and forms of resistance. In R. L. Harris & J. Nef (Eds.), *Capital, Power, and Inequality in Latin America and the Caribbean* (pp. 196–224). Lanham, MD: Rowman & Littlefield Publishing Group.

Law 200 of 1936 – About the Land Regime (1936).

Law 100 of 1944 – About the Land Regime (1944).

Law 99 of 1993 – Creates the Ministry of Environment and the National Environmental System (1993).

ONIC (National Indigenous Organisation of Colombia). (2013). *Mandato Político General del VIII Congreso Nacional de los Pueblos Indígenas de la ONIC 2012–2016.* Bogotá, Colombia: ONIC.

Pereira, A., & Rappaport, J. (2022). *The participatory research of Orlando Fals Borda.* Brighton, UK: Participation Research Cluster, Institute of Development Studies. Retrieved from www.participatorymethods.org/resource/participatory-research-orlando-fals-borda

Perreault, T. (2006). From the Guerra Del Agua to the Guerra Del Gas: Resource governance, neoliberalism and popular protest in Bolivia. *Antipode*, 38(1), 150–172. doi:10.1111/j.0066-4812.2006.00569.x.

Political Constitution of Colombia. (1991).

Pratt, B. (2013). El proyecto de gas de Camisea: los movimientos sociales indígenas y las ONG internacionales en la Amazonía peruana. In A. Bebbington (Ed.), *Industrias extractivas, conflicto social y dinámicas institucionales en al Región Andina* (pp. 279–300). Lima: IEP, Instituto de Estudios Peruanos.

Rappaport, J. (2000). *La Politica de la Memoria.* Popayan, Colombia: Editorial Universidad del Cauca.

Rappaport, J. (2020). *Cowards don't make history.* Durham, NC: Duke University Press.

Rival, L. (2013). La planificación de los futuros desarrollos en la Amazonía ecuatoriana: la frontera petrolera en expansion y la iniciativa Yasuní-ITT. In A. Bebbington (Ed.), *Industrias extractivas, conflicto social y dinámicas institucionales en al Región Andina* (pp. 249–278). Lima: IEP, Instituto de Estudios Peruanos.

Sieder, R. (2007). The judiciary and indigenous rights in Guatemala. *International Journal of Constitutional Law*, 5(2), 211–241. doi:10.1093/icon/mom007.

United Nations Declaration on the Rights of Indigenous Peoples (UNDRIP). (2007). 61/295 C.F.R.

Velasco, M. (2011). Contested territoriality: Ethnic challenges to Colombia's territorial regimes. *Bulletin of Latin American Research*, 30(2), 213–228. doi:10.1111/j.1470-9856.2010.00500.x.

Warnaars, X. S., & Bebbington, A. (2014). Negotiable differences? Conflicts over mining and development in South East Ecuador. In E. Gilberthorpe & G. Hilson (Eds.), *Natural resource extraction and indigenous livelihoods: Development challenges in an era of globalization* (pp. 109–128). Surrey, Burlington: Ashgate Publishing Ltd.

6 Responses to Mining in the Nasa Indigenous Resguardo of Canoas

Introduction

It is in the Resguardo context that one can best appreciate the tensions and difficulties that Nasa communities encounter with the arrival of mining in their territories. The Resguardo domain provides a space to explore Nasa understandings of mining in a highly contextualised way. The stories cease to be general and become about neighbours and places or periods of shared community life. They gain more meaning in this localised context and with the specific local politics of mining as a framework there is ease of interpretation. While one Resguardo is very different from the next, it nevertheless illustrates the Nasa Indigenous movement's positions and tensions in a unique context that holds commonalities with other Resguardos and can also illustrate the contrasts.

Here, I shall examine Nasa understandings and responses to mining within the local context of the Nasa Indigenous Resguardo of Canoas. The choice of Canoas for the case study was the result of three factors: i) a judgement of relevance by ACIN senior people that placed Canoas among the Resguardos to consider for further examination; ii) the Canoas political process and leadership decisions on the priority that a study about mining warranted; and iii) my evaluation of levels of community organisation and support for research in Canoas, as well as readiness within the timelines of the fieldwork, security and logistical considerations. In the Introduction to this book, I narrate the process that led us to selecting Canoas as a case study.

Here, I concentrate on the perspective of opposition to mining that was conveyed to me by several Canoas authorities, leaders, researchers and community members and on the difficulties and tensions those seeking to control mining activity have faced. The argument I shall present is that, despite multiple concerns about mining and efforts to deter it, material scarcity in Canoas has made mining difficult to control. Where Canoas has often been the site of innovation across the ACIN affiliated Resguardos, its responses have not resolved the mining conflict. The disagreements between miners and those against mining have included periods of intense conflict and tense calm with a solution of remaining elusive. The community has found itself in a cycle of 'command and

DOI: 10.4324/9781003226895-6

control' measures by the Authorities followed by reversals or retaliation from the miners. Most participants who were not miners saw a halt to the activity supplemented with livelihood alternatives as pre-conditions for peace. They were concerned about limited livelihood options, poor mining regulation at the national level, the moral degradation linked to mining, and the pervasive links between mining and the armed conflict that hindered peace. These concerns lead me to stress the need for livelihood alternatives and cross-institutional, cross-cultural collaboration in creating possibilities for a peaceful future that includes demobilised armed group militants.

The testimonies that this chapter is based on come from 17 Canoas residents whom I refer to by pseudonyms of their choice. In some cases, these pseudonyms are in NasaYuwe. The purpose of the pseudonyms was protecting the identities of the participants. There was no intention to analyse the meanings behind the participants' choices. The purpose was to refer to the participants with names that are meaningful to, or at least chosen by, them while preventing identification.

I present the analysis and argument in four sections, followed by a brief discussion. The first section, this introduction, will conclude with a description of the Canoas Resguardo where I seek to situate the reader in the local context. The second section is about mining in Canoas. I provide a discussion about mining in the Resguardo from a legal and chronological perspective, and offer some insight into why there is mining there. Third is a discussion about experiences and concerns with mining in Canoas. Fourth, I outline Canoas's past responses to mining and the plans and hopes for a future response.

The Canoas Resguardo

The Canoas Resguardo is in the rural area of Santander de Quilichao, in North Cauca, and consists of 27 villages on the western front of Colombia's central Andean mountain chain (ACIN, 2013; Canoas Cabildo (Canoas), 2017). It had a population of more than 9,300 people, or over 2,700 families. Forty-nine per cent of the population was female (Canoas Cabildo (Canoas), 2017). The people of Canoas represented less than 8.3 per cent of the Nasa population of North Cauca. Canoas is a highly organised Resguardo, connected to the Indigenous organisations of the region. It has a strong agricultural identity. Like other Nasa Resguardos, it has faced hardship because of the armed conflict and material poverty.

The Canoas's Cabildo is affiliated to ACIN and CRIC. At the time of the study, the Canoas Cabildo, which is the Indigenous Authority of the Resguardo, comprised 35 people or Authorities. The Assembly is the most important Cabildo participation space, where 2,000 to 3,000 community members take part, three times a year. The Assembly elects Cabildo members yearly. As with previous Cabildos, the 2017 one had economic-environmental, health, and child-care programs; as well as the Guardia and a harmonisation centre, where people who have committed offences are re-harmonised or reintegrated. The Cabildo and Guardia members are volunteers in these demanding full-time roles. Besides this, the community receives earmarked National Government fund transfers for specific programs. The Cabildo seeks further funding from international aid organisations.

Water is a central aspect of Canoas's identity. Clusters of ACIN affiliated Resguardos produce Life Plans that outline the communities' aspirations and plans. Canoas is part of the Yu´Lucx Life Plan, together with the Munchique and Nasa Kiwe Teckshw Resguardos, also from Santander de Quilichao. Created in 1991, Yu'Lucx translates as Sons of Water (ACIN, n.d.). The creation story tells how the first Nasa ancestor is a descendant of the Star and the Lagoon. The Resguardos within this Life Plan harbour important water supplies. Water sources in Canoas provide water for 27 municipal water tanks in the Santander de Quilichao municipality. The Resguardo negotiates water use and management on an ongoing basis with campesino communities (Canoas Cabildo (Canoas), 2017).

Canoas has a primarily agricultural economy and is said to be particularly productive compared to other Nasa Resguardos. The community land titles cover 1,200 hectares. A Canoas resident, Seklucx (pseudonym), described the economic life at Canoas thus: "*People grow green beans, pineapple, coffee, plantain, kidney beans, cassava, chooks, they build lakes and grow fish. You don't make a lot of money, but you live happily.*"

Canoas has a tradition of strong environmental campaigns and its members seek to highlight the Resguardo's and, in general, the ethnic people's contribution to wider society through environmental protection. Rather than 'anti-development', as they are often labelled, the Nasa of Canoas see themselves as guardians of nature. Seklucx explained it this way:

> *This river is not just for the Indigenous communities, this goes down to the valley, so there are the Afro-descendant communities, there's the campesinos, there's the entrepreneurs of the Cauca Valley. So there is a chain that we need to look into. So that is our objective, to demonstrate that the Indigenous people are not anti-development; on the contrary we are protecting the environment because when the trees produce oxygen it is not just for the Indigenous people but for everyone. […] Unfortunately, the multinationals are only looking at the money sign, and they don't look at what we are protecting.*

The predominant understanding is that this role of Indigenous communities is not valued, and that other economic forces relentlessly seek to dispossess them of their lands based on ideological arguments about the economy. Seklucx continued:

> *The Government said that the Indigenous people were slackers because we had the mountains and were not working them. He looks at one thing but we look at something very different, because by preserving the mountains*

> *we are strengthening oxygen, water resources, protecting the large gemstones that are like the veins of our Mother Earth. [...] But external organisations like the multinational mining companies have always sought to displace us. So have the war and the threats.*

Despite this environmental protection and agricultural tradition, several social, political and economic factors are putting pressures on livelihoods and the environment. The population is growing at a fast pace and the Resguardo has diagnosed land scarcity. Canoas is the most "*land tight*" of the Nasa Resguardos (ACIN, 2011). Nasawala (pseudonym) told me that the most fortunate have two hectares, while some have less than a quarter hectare, and a growing population is seeing more young people seek access to land. According to Nasawala, Canoas's traditional agriculture has made it a key provider to the Santander de Quilichao area. However, oversupply and the declining prices, together with the land availability problem, push people into other economic activities such as mining that cause environmental degradation and conflicts.

Canoas has faced, and continues to face, difficulties because of the violence and armed conflict, and illicit economies have managed to take hold during some periods. Ul (pseudonym) told me about the waves of violence against leaders in the 1970s, and the 1980s drug trade related violence. Later in 2006, "*Mining arrived with the same violence of crops of illicit use*", he assured me. "*However, we resist. There are always external people coming with things and with violence. We have a very good organisation in our Cabildo. If it wasn't for the Cabildo we would be worse off. But the young people, they have gone somewhere else. If they become miners or grow crops of illicit use, we have no funds to help them. The poverty here is tremendous*". The confluence of the armed conflict with material poverty has created significant difficulties for the Canoas community.

Mining in Canoas

In this section, I shall discuss the Indigenous laws and mandates about mining that are in force in Canoas. A brief description of how the mining dynamics have evolved over the years in Canoas will follow. Later, I examine why it is that in Canoas, an Indigenous territory that banned mining, there were still Indigenous miners.

Canoas Law and Mandates about Mining

In the Nasa cosmogony, gold is associated with power and protection. This source of power can only be managed with a great deal of care and should not be extinguished. Guia (pseudonym) narrated to me the story about The Thunder (*El Trueno*) and its relationship with gold. He described The Thunder to me as a set of spirits that have defended the Nasa since the ancestral wars with their Indigenous neighbours, the Pijao. The Thunder represents the strengths of the Nasa and gold must remain in the sacred mountain of Cerro Munchique so that The Thunder can protect the Nasa. Guia told me:

> *Gold belongs to the Earth. It belongs to The Thunder. It wasn't ours. If there is good gold, if the mountains have sufficient gold, there is more connection of The Thunder with the Earth. So, if The Thunder leaves, there will be no defence anymore. We depend on The Thunder. For example there is a myth. […] In the first war that the Nasa had, they fought the Pijaos. So because the Pijaos ate the Nasa, then The Thunder, the owner of the gold, told an Elder that he needed two orphan boys. So he received them and turned them into Thunders as well. So there were three Thunders. One was called The Owner of the Axe. The other was the Owner of the Sling, the one used to throw stones. So the Nasa won the war. And the orphan boys were no longer Nasa but Thunder. They could not set foot on the Earth anymore. So the Thunders were absent, and disorder started again. So the three Thunders thought about a fourth. The fourth Thunder is The Owner of the Chonta [or the baton] of the Traditional Doctor who consults the other Thunders to know what to do. So if they devastate the mountain then the Thunders will leave and there will be no protection.*

Both the rivers and the sacred mountain of Cerro Munchique are essential sources of life to the Nasa, and it is precisely in these places where the miners have been. This creates fear that mining can extinguish life. In Guia's words, "*If they devastate the Cerro [Munchique] they will devastate our life, because it is everything, it is like a lung for us*". Furthermore, the Nasa do not accept the idea of mining for accumulation purposes. A participant who chose the pseudonym 'Mayor' told me:

> *There is no word in NasaYuwe. Mining is like a resource, like extracting it. The idea of resource leads us to think about money, and money, from the perspective of Western culture is living comfortably. But instead of mining we talk about looking after the territory, about looking after the source of life.*

The idea of mining precisely in that source of life and strength is disapproved of and feared.

Canoas has been active in environmental conservation through territorial organisation, environmental education, research and evaluation, and this is reflected in its environmental rules. The Canoas Indigenous Reserve Environmental Rules (2015) aim to provide "*mechanisms to help recuperate and preserve natural resources and mitigate the negative impacts that might alter [their] relationship with Mother Earth and the environment*". The Environmental Rules refer to the water, air, forest and soil impacts of various activities including mining. They speak about the exploitation of minerals without a prior study and without mitigation actions, about mining's water use, its use of chemicals, and the deforestation it has caused. Based on these, article 26 prohibits mining in communal or private lands where "*failure to comply will lead to suspension of the activity, closure of the mine and confiscation of machinery*", as well as penalties. The Rules give responsibility to the agro-environmental committee of the Resguardo to coordinate compliance and liaise internally and externally to that end.

The debate to arrive at this mandate was long-standing and nuanced. The records of the first Canoas Congress in 2010 make clear that the community was coming to terms with "*the mining problem*" and the Congress called for an environmental impact study about mining in the Resguardo, and for clarity as to the conditions within which mining could take place. It is important to bear in mind that gold is only one mineral, and there are other vital minerals that the community needs, such as construction materials. The Congress conclusions state that the Resguardo was not to allow externals, such as multinational companies, to mine in the territory and that there should be no mining without an environmental impact study (EIS) being conducted first so that the Resguardo could make decisions. The questions the community was working on were about the legal framework of mining in Colombia, the environmental impacts of mining and what kind of mining could be permissible. Within five years, however, the growth of this activity and its repercussions in the community resulted in a very clear mandate of NO mining within the Resguardo.

Mining in the Canoas Resguardo

The mining dynamics in the Canoas Resguardo have undergone significant shifts over the last four decades. Table 6.1 shows an approximate timeline of some of these changes as seen through the eyes of the people of Canoas. This is based on documents from the Canoas Cabildo and testimonies of the research participants. The timeline seeks to illustrate the shifts as understood by the local people rather than provide precise dates or event descriptions. Most of the timeline data I received was in terms of periods of several years at a time rather than specific dates. Many of these events relate to illegal activity linked to armed groups that is difficult to corroborate. Nevertheless, this brief timeline illustrates how people have observed conflict gradually escalating, particularly following the 2007–2008 period of favourable international gold prices.

Mining triggered social conflict and periods of intensified armed conflict. The prolonged presence of mining in the Resguardo resulted in a protracted conflict and a degree of polarisation that makes it difficult to obtain accurate information to characterise mining. However, the overall perception was that the number of people working in mining in 2016 and 2017 was lower than it was a decade before. The number of miners or mining tunnels and the production levels were not known. A miners' census was yet to be conducted and

Table 6.1 An Approximate Timeline of Mining in Canoas

Pre 1970	Non-Indigenous landowners had mines in what is now the Canoas Resguardo and used Afro-descendant and Indigenous people as "slaves" (Canoas Cabildo (Canoas), 2017).
1970s	The Canoas community concentrates on dairy and coffee production.
	Mining tunnels remain open at the sacred mountain of Cerro Munchique.
1992–1993	A local begins to mine in one of the tunnels in an 'artisanal' way for two or three days a week to supplement income from blackberry production. The Cabildo allows it.
	A man from Antioquia known as El Paisa brings machinery and personnel to work in one of the tunnels of Cerro Munchique.
	Community complaints about El Paisa's activities' effects on water lead the Cabildo to confiscate the machinery and agree that El Paisa should leave.
1996	The Canoas and Delicias Resguardos obtain declaration of an Indigenous Mining Zone.
2008	A multinational mining company applies for priority rights for gold exploration and exploitation in Canoas (Caro Galvis & Valencia, 2012). It offers the Cabildo livelihood and infrastructure projects, as well as funds.
	The Cabildo refuses the multinational's request and offer and "*makes it clear that it will resist multinationals' entry*" (Canoas Cabildo (Canoas), 2017; Caro Galvis & Valencia, 2012)
	The Cabildo faces pressure from the Government (Santander de Quilichao Mayor's Office and Ministry of Mines and Energy) (Caro Galvis & Valencia, 2012) but ACIN helps resolve it.
	Some residents of El Condor village manifest their interest in mining gold in the Cerro Munchique "*lest multinationals or externals do it*".
2008–2009	Illegal alluvial gold mining using machinery begins in the Mondomo River.
	A mining leader from Canoas is found dead.
2010–2011	Alluvial mining re-activates in the Mondomo River.
	The Canoas Congress holds a debate about mining: No mining allowed to outsiders. EIS about mining in Canoas to be conducted.
	Six months of daily army-guerrilla combat begin at the Mondomo River as army attempts to stop the miners while guerrillas protect them.
	Approximately 300–400 local Indigenous people are found working in the Mondomo illegal operation.
	The Cabildo, Guardia, community and students expel the illegal miners from the Mondomo River and there are confrontations.
2011–2012	Some residents of El Condor re-open a mining tunnel in the Cerro Munchique.
	The El Condor mining affects the Pacadoa aqueduct that services six villages in Canoas.
	Residents of the Paez village and a further five villages in the Resguardo complain to the Cabildo about water impacts from the El Condor underground mining.
	It is said that 200–300 Canoas residents work in the Cerro Munchique tunnels.
	In Assembly, the Cabildo and miners agree that the miners will stop for a month while the Cabildo seeks financing for agricultural projects and for a community fund to assist the transition away from mining.
	El Condor miners return to the tunnel before the month is up.

2012–2013	The Cabildo, Guardia and some community members go to El Condor to close the tunnel and this marks the beginning of a series of working bees against mining, often in collaboration with ACIN and other social organisations.
	The next day El Condor miners re-open the tunnel.
	A new Cabildo Assembly is programmed.
	The miners begin a campaign against the Cabildo.
2015	The Canoas Cabildo updates its Environmental Rules and re-states a ban on mining.
2013–2016	Subsequent working bees against mining are implemented and a dynamic of "*blocking and un-blocking tunnels*" perpetuates as miners and the rest of the community remain in conflict.
2017	A Canoas miner is murdered in Munchique when a group is found trespassing in a gold mining tunnel on a private property.
	A series of working bees to stop mining are implemented. The North Cauca and Cauca Indigenous Organisations get involved and national and international observers attend.

the estimates of the numbers varied widely. Most participants stated that the miners were a minority, however the miners disputed this.

Community observations refer to two types of mining in the Resguardo as of 2017: underground mining and alluvial mining. Underground mining happens in the slopes of the Cerro Munchique Mountain, often in old tunnels that were in use before the 1970s. There are several mills in the Resguardo to process the soil. The miners I spoke to assured me that the mills within the Resguardo use only mechanical rather than chemical separation processes. Reportedly, the soil by-product is sold for chemical processing outside the Resguardo. As for alluvial mining, some people use dredges in the riverbeds and others pan for gold.

Canoas people are suspicious of the source of financial backing for artisanal mining. In Seklucx's words, "*Some miners did not have money and suddenly they have drills to go and make holes under the land*". The Canoas participants echoed the concerns of people from ACIN and from other Resguardos who were convinced that external money was fuelling illegal mining in Indigenous territories. Some believed it could be a formal mining company that, unable to gain entry to the Resguardo, had decided to purchase the illegal gold, while others believed it was illegal armed groups or a mix of both.

Why Are There Indigenous Miners in Canoas?

Most non-miners observed that young men are strongly represented among the miners and some believed this is the result of economic trends that have not been appropriately managed. These men have young families and dropped out of school to support them. A Nasa is considered an adult at 15 and at that age might be looking for a partner to start a family. A participant who chose the pseudonym 'A te' told me that some of these young men have limited resources and, used to being in the countryside, find in mining a suitable, or even the only, way to make a living. There had been instances of mining before and there is much difficulty in finding an appropriate response because of the scarce funds and because mining is such a polarising problem. People are likewise divided when it comes to judging Cabildo actions on the matter. A te is among those who believe the Cabildo, and the State, lacked foresight to prevent the lack of economic opportunities for young inhabitants of Canoas. Some local leaders interested in working with youth had formulated projects to fund agricultural work, but these had not received Cabildo or external support, according to Ul (pseudonym), an experienced agriculturalist and environmental advocate.

The situation is difficult to reverse because once young people become miners it is unlikely that they will want to quit. A te told me the miners can make in one night what they make in one month as seasonal agricultural workers; while Pil (pseudonym) had been informed that with two or three grams of gold that people might be able to find on a given day, the day's earnings ascend to COP $120,000, whereas a day of agricultural labour pays COP $20–25,000. Other people had given Pil examples of miners who had made around COP $7 million in one month. To put this in perspective, the minimum wage in Colombia was just over COP $781,000 at the time of the research. The financial balance tilted conclusively towards mining.

Nevertheless, several participants thought what attracted people to mining was spiritual disharmony rather than material need and that the benefits of mining are elusive in any case. Agata, Ortensia (pseudonyms) and others observed that mining moneys were used for destructive activities, not to build economic prosperity. To Agata, this was proof that it was disharmony that led people to mining to begin with. She said that, "*When you are suffering disharmony you don't see the territory with respect but as something you can utilise for individual gain. […] We have seen the money is not put to good use. It is wasted in alcohol and in things that have brought disharmony to their families […] If they had clarity on why they are doing mining, they would concentrate on other things*". So mining money brings disputes to the communities. Furthermore, the material benefits are limited. In Ortensia's words, "*Mining doesn't show; very few people have bought a little house; what little they make they drink it, it is for alcohol; so you can also see the damage, except for a car or a motorcycle here and there. What happens is that whoever gets money starts drinking, finds a new woman and leaves his partner with the children at home. I don't see any benefit*". Almost every participant, aside from miners, described similar dynamics.

Lived Experiences of Mining in Canoas

The experiences that participants recounted about mining in Canoas encapsulate three main concerns of those opposing mining. First, that formal mining is poorly regulated and puts Indigenous rights at risk. Second, that something negative happens to people when they 'touch' gold; they deteriorate and drag the community down with them. Third, that there are tangible direct and indirect links between mining and the armed conflict, as well as a very steep path ahead to build peace in territories affected by illegal mining. I discuss each of these in turn.

Poorly Regulated Formal Mining: A Risk to Indigenous Rights

Just as many in the Canoas communities rejected local or external miners conducting unsanctioned mining in the Resguardo, all those who spoke to me were very worried about formal mining. The concern about mining in Canoas applied to formal mining companies and enterprises and to what was commonly referred as 'the multinationals'. Canoas residents had observed, and Colombian research has confirmed, that institutional weaknesses affect Colombian regulation of formal mining. Mayor explained it thus, "*We are greatly concerned about the proposal to implement the mining locomotive. We know that legally things are not strong for regulation, and environmentally there are no teeth to regulate and minimise the impacts*".

In that environment of institutional weakness, Mayor believed the rights of Indigenous people were violated on a day-to-day basis to promote mining in their territories. Formal mining, to Mayor, was a force driving a bias in the law to the detriment of Indigenous rights and local priorities. He said:

> *The 1991 Constitution – our first peace agreement, that recognised self-government and territory and gives us the right to prior consultation before any project is implemented in our territories – that right's been violated almost every day in Colombia because the rules and regulations are not made to accommodate difference, that is, considering our way of thinking, but rather the interests and economic needs of transnational corporations and the macroeconomic interests that globalisation imposes on us.*

Furthermore, there was a widespread perception that 'multinationals', 'transnationals' or 'foreign interests' were linked to unauthorised mining. Many believed there was a causal link between formal mining company interest in a territory and the arrival of unauthorised miners. Some believed the link was through direct sponsoring. Others thought once a company showed interest, unauthorised miners took that as an indication of gold presence.

When People Touch Gold: Mining and Disharmony

Agata, A te, Alegria, La Profe, Ortensia and Luz Dary (pseudonyms) all associated mining with the arrival of disharmony to the Canoas community. The female participants described it as something that happens to people when they 'touch gold'. Something changes in the person. They become thirsty for more gold. Their ambitions grow as access to cash makes them familiar with consumption. Regardless of the rewards, the women explained, the person will continue, desperately seeking gold at the expense of their wellbeing and that of others. All participants saw family dysfunction as a direct result of participation in mining. Male participants disapproved of the sumptuary consumption that miners engaged in because they saw it as a form of status signalling aimed at humiliating fellow community members, particularly males. They believed cash availability enabled substance abuse and access to weapons, which were both used to assert status among males.

For most, the mere knowledge that gold was under the land has become a problem at the Canoas and Munchique Resguardos. It attracted external interests to the Indigenous territories that set a chain of reaction in motion. Guia recalled that, "*There was a study [...] before that nobody knew about the gold [...] now people know and they say 'we'll do it before the multinationals come' [...] so they don't worry about damaging the land*".

Once people come in to contact with gold, that interaction seems to change the way they think. To A te "*There are certain forces there*" that make people think like this: "*I have gold, I have power, I can have things that the rest cannot, I am a miner, I have money, I have status*". She stated "*it is difficult for them to change because they have already touched money. They know that you can do many things with money*". La Profe described how people's way of thinking and feeling changes when they become miners. Their wealth, or the promise of it, makes them lose trust in their fellow community members. "*To me, when I don't know gold*", said la Profe, "*then I hold that value of preserving nature, of believing in spiritual life and of believing that the Earth is our mother*". But she added that:

> *When you start touching gold, or to exploit it or to have, then you forget all that, 'no Mother Earth nonsense or anything like that, what do I need now? I don't have any job alternatives and I am in need'. Even if they have other alternatives in a company, that money takes time to come, while gold is quick. And I say that changes people, because say I am getting gold and I can see that I am doing well, then I cannot even trust myself. That is when you decide to buy a weapon, because at any time they might come to steal from you [...] It damages people's thinking.*

According to these explanations, this way, community members are gradually lost to mining. They connect with gold and that relationship comes at the expense of others because the greed for gold erodes trust.

Such a dangerous relationship with gold resembles what the Nasa's ancestors and Elders experienced. Ortensia explained to me that mining had a place in the Nasa ontology and cosmogony, but there was an awareness of the dangerous qualities of gold. The tradition was to respect gold, to work it with aesthetic and ritual purposes, to be careful and to avoid excess through discipline. Ortensia said, "*People would work the gold and make their little things and jewellery. Then the Spanish came for the gold and killed everyone and left devastation in their path*". She recalled the conversations from the time an Indigenous burial was found. Ortensia said, "*They do rituals with that. People say that gold is very jealous with people. If you have ambitions to find gold, you won't find it*". She said gold does not reward those who search it with despair and gave me an example of an acquaintance who is ambitious and searches tirelessly, but finds nothing.

That danger of becoming enslaved by the desire for gold led the Nasa ancestors to relate to it through discipline and ritual practices, but instead, the contemporary miners were immersed in a cycle of ambition and frustration with gold at the centre. The traditional rituals and discipline that characterised the Nasa's relationship with gold ensured the Nasa could benefit from gold over a long period of time and protected them from becoming dominated by its power or victims of those afflicted by the greed. Nowadays, however, miners' relationship with gold is different. Ortensia and several other people recalled relatives who knew where to get gold, would go and carve a small piece to sell in times of need, without using it up or telling others where it was. Frustrated, Ortensia said to me, "*It has been a long time since gold has existed and that is how people worked it. But the thing now is that if there's gold they HAVE TO get it out and they HAVE TO go under the earth, because they start with their peaks and they HAVE TO follow the vein*". She was appalled that others thus were compelled.

Most Canoas participants were very concerned that where a family member, usually the father, takes up mining he goes through negative changes that affect his entire family. Increasingly he uses the mining income for alcohol consumption and abuse. He begins having affairs and neglecting his family responsibilities. La Profe explained that, "*They don't pay attention to whether the child is well, or needs to be taken for a walk, or how the child is doing at school, or the child's nutrition*". She reflected that before there was gold those families had a balanced diet, as opposed to when they make money from gold because "*you cannot really see positive outcomes*". "*I think you can see that there is social chaos and family problems*", she added. La Profe, Pil, Sek and others observed that mining leads not only to child neglect but also to school desertion (Caro Galvis & Valencia, 2012). In 2014, there was a wave of school truancy that they attributed to miners leaving their children to fend for themselves. La Profe was relieved that this had stopped following sanctions against the parents.

The problems quickly spill beyond the family through what male participants called "*humiliation*". According to Seklucx, Paez and Enrique, miners do not know how to use large amounts of cash other than for consumption of sumptuary goods to assert male status by "*humiliating*" their peers. It was mostly the male participants that spoke about this form of humiliation. They described it as an appalling practice based on pride and selfishness. Whoever can buy more whiskey or have the most expensive motorbike can humiliate others through those acts of consumption that people with lower incomes cannot afford. These men take the rivalry as far as competing for each other's wives. Guia told me about the spiritual deterioration that leads the miners to such behaviours. He explained that each person has a good and a bad spirit inside and they live in tension. When something triggers pride and vanity, or the bad side, the Nasa person might not be prepared to deal with that. To Guia, "*If money comes to you and you are not prepared, money will kill you; it is like sitting behind the wheel of a car without knowing how to drive,*" he explained. I asked if this happened with all miners. Guia told me that all people have a conscientious side and one that is not so. "*From a young age, we are forbidden to eat in front of other people if we are not going to share*" he emphasised, and "*that who is conscientious knows the needs of others,*" Guia added. Those people who are conscientious can see that they are hurting others. They can observe their behaviour, seek help from the Traditional Doctor and do the rituals to regain harmony. However, Guia told me those who are not conscientious "*only think about themselves*". As Agata affirmed, the moral transgressions of the miners are attributed in part to their spiritual problems.

Canoas was not exempt from the divisive dynamics many attribute to mining. The division had gradually escalated. Seklucx and Paez described a process where conflict encroached into larger and larger domains. The substance abuse and family neglect erode the family first. Later, numerous complaints and disputes – be it about water and Mother Earth, humiliation, or infidelities – set fellow community members against one another. For example, there had been clashes between the miners and the Canoas Guardia supported by the community. The challenges to the Cabildo followed. The village of Jerusalem has mining and is linked to a dissident Cauca Indigenous organisation that contests the authority of the Canoas and Caloto Resguardos. Ortensia recalled that after a confrontation with the miners at El Arbolito village, some said they were tired of the Canoas Cabildo and would start a new one. There has been a fatal confrontation between Canoas and Munchique miners. Furthermore, the disputes about mining saw the two Resguardos face tensions. What began as smaller discrepancies grew until more and more people were involved in the conflict or seeking to appease it.

People observed strong links between mining and violence. There were links to the armed conflict that I shall discuss later, and a number of moments of crude violence that the participants remembered with concern and sadness. Luz Dary and Sek (pseudonym) told me about an armed robbery connected to mining that resulted in three deaths. It was perpetrated against illegal miners from outside Canoas who were working there. It had happened some years back. The group of miners were shot at when driving out of Canoas. This led the car off the road and all occupants were killed. A Guardia member found out that, in Munchique, there had been other deaths in machete fights related to mining before the 2017 murder. According to Sek, there had also been unexplained deaths during the illegal mining boom of Canoas.

Not only were the benefits elusive, but the problems resembled a scenario that the Nasa feared repeating: the disharmony that to them led to the Paez river avalanche, a natural disaster in 1994 that changed the lives of thousands of Nasa from the Tierra Adentro area in Cauca. Guia saw worrying similarities between the mining-influenced communities and Tierra Adentro in the early 1990s. The Nasa believe the avalanche was Mother Earth's reprimand and warning, because they had been growing poppy, which is used for heroin production. Guia drew multiple parallels between the violence and social deterioration that happened as a result of illicit poppy cultivation, and what happens with mining:

> *For example, last week, the person who died, what killed him was a shotgun wound. Once that happens, what can the Cabildo do? This is going to be like what happened with poppy. It was so expensive so people armed themselves. The Cabildo could do nothing. People were killing each other and painted the houses with blood. Then the Earth trembled. The Elders said that was no punishment but a warning about the damage we were doing to the land, and to ourselves. Now we are going down the same path. There will be no Cabildo that can stop this. They'll organise themselves. They can name Gobernadores. Money can do anything.*

Like Guia, many Nasa participants were concerned that the environmental and community deterioration would lead to a response from Mother Earth, that a natural disaster would come as a result of the trespasses and all would pay for the mistakes of the miners who disobeyed community mandates.

Mining, Armed Conflict and Peace

I now examine how the Nasa of Canoas understand the interactions between mining, armed conflict and peace. Let us begin with the armed conflict. The links between mining and the armed conflict had been very clear in Canoas while the conflict with FARC was still ongoing, but when talking about the post-agreement transition, the participants found the links difficult to describe. Violence was still there, but in whose hands? There was a sense that as violent groups reconfigured, some of them enjoyed a newfound anonymity. Nevertheless, participants observed armed people gravitating around mining and several felt confident to identify them as guerrillas, criminal bands or other groups both legal and illegal.

During the armed conflict period, participants observed strong causal links and interactions between mining and the armed confrontation. These included mining-motivated FARC-Army combat within the Resguardo; miners' neglected children becoming easy recruits for FARC; and guerrillas, army and paramilitaries extorting payments from illegal mining. Remembering the time when FARC were active, A te described a direct link between mining and the armed conflict:

> *It began in 2010. That area has always been very quiet. The hostilities between paramilitaries and guerrillas were further downhill. That year, they*

started excavating the river. You could see the army come up. Then the guerrilla would come up. It was like clockwork. Every day, at around five pm, the combat would begin. It was very consequential: mining begun, the army arrived, the guerrilla faced them and there was combat. [...] I don't recall very well but I believe it lasted six months. It only stopped when the Cabildo decided to expel those people [miners].

Canoas residents witnessed armed groups, legal and illegal, extorting illegal mining operations. Mayor explained to me that the links were observable in the territory. "*In the Palo river, there is mining too*", he lamented. He explained, "*You could see how every day at four pm [FARC] militants would go down to receive extortion payments. You could see strangers in black uniforms with ranged weapons, who we inferred were paramilitaries, coming to look after the machinery; and the army, only two kilometres away, set up the check point and that business went through there*". Mayor told me that common crime increased and later the murders of Indigenous, campesino and Afro-descendant leaders started. They were killed "*simply for having denounced the situation*", he said. In the Mondomo river it was no different, according to Mayor, "*The guerrillas provide security, then the paramilitaries provide the transport service from there to Jamundí and Cali*". Seklucx concurred and assured me that the local Guardia had seen enough evidence that both legal and illegal armed groups were taking a cut in illegal mining revenues. There were similar narratives about San Antonio, in Santander de Quilichao, where dozens of people have died in illegal mines after wall collapses. Machinery could have only reached the site through the heavily militarised and transited Panamericana road, but "*nobody saw anything, they just did the rounds*," said Mayor. One of these sites, Mondomo, is in the boundary of the Resguardo, and the situation had changed, Mayor explained, only after the Cabildo took action.

During the transition period, as armed groups were reconfiguring, the culprits were harder to identify but similar forms of violence, linked to mining, remained. Agata noted that the Nasa knew there were armed groups connected to mining but they were hard to name. Ortensia was not sure whether there were armed groups connected to mining in Canoas but believed it was different in other Resguardos. Luz Dary recalled that during a visit to El Condor village to talk to the miners the Guardia were shot at, but she was not sure who fired the shots. Seklucx believed it was the miners who were arming themselves. Alegria noted with concern that the situation during the post-conflict transition was worse than in 2011, because back then there had been no murders among the community. However, there had been ongoing threats against Authorities. Another concern for many was whether FARC ex-combatants would join the ranks of the miners. A te worried that former combatants might find themselves without other sources of income, where mining was one of the most akin occupations for people used to 'bush life'.

How does mining interact with peace? The message from participants, other than the miners, was that to build peace, alternatives to mining needed to be found. The logic for this was threefold: i) the spiritual, social, political, economic and environmental shifts that mining triggers are an obstacle to peace; this is because ii) peace is Living Well or *Wët wët fizenxi* and mining conflicts with *Wët wët fizenxi* and with the requirements of peace[1]; as a result iii) it is necessary to stop mining and find alternative livelihoods to be able to build peace. To most participants, it was clear that economic and spiritual precursors precipitated involvement in mining so peace could only be built with holistic responses to those problems.

Guia explained to me that the spiritual transformations that mining caused in people were an obstacle to peace because the miners ceased observing the Origin Law of the Nasa. The Nasa talk about peace and moral ideals with reference to an ancestral past that provides a template for the ideal life. Guia explained:

> *In order to build peace we must look into the past. In the past people had laws. We call them Origin Law. They are: visit each other, help each other and maintain dialogue with each other. If you had a problem you'd talk. People would cook for others, help build their houses, bring firewood when a child was born [...] If you share, if I live sharing, people will share with me, but if I don't then they won't either. In order to build peace we must look into the past. But if we only look into the future then how are we going to build peace?*

He suggested the self-interest of miners and their urge to humiliate was sending them down a path too distant from what Origin Law prescribed and to him it was that Law that maintained peace in the past.

The Nasa understanding of peace is based on their cosmogony: they equate peace to Living Well. Mayor's discussions about peace align with concepts of positive peace, as was the case with many other participants. His discussion exemplifies how peace relates to some of the Nasa principles and ideals and, in light of the mining-induced disharmony discussions earlier, illustrates the obstacles that mining can pose for peace. Mayor said:

> *Sometimes peace is thought of as a warranty of wellbeing for the community or an end to the war. We need to revise that. More so for us because we have suffered sixty years of war and, in particular, the Indigenous people who have suffered 540 years of war. Peace for us transcends those concepts and definitions. Saying that there must be peace is saying that there must be* Wët wët fizenxi, *or Living Well, living beautifully, happily in the territory.*

Mayor told me there should be a healing process to attain territorial and spiritual peace and that peace for the Indigenous people is contingent on the peace of the Colombian people. Throughout his discussion, he highlighted reconciliation with the Nasa playing a fundamental role, in particular in healing the territory. Called to be active peacebuilders, the Nasa cannot but attend. Mayor told me:

> *We must follow a process to heal ourselves, in the heart and in the mind. It has to come from our ritual life. We must heal the Earth as well – because blood has damaged it, the bombs have damaged it, the explosive artefacts, the landmines. So that the Earth can go back to giving us like she gave us before. But peace means healing the large social fabric that the conflict broke. In this country, that's what we need, so that we can speak again like brothers and sisters. That's why there cannot be peace for the Indigenous people if there is no peace for the Colombian people. There can be NO peace for the Indigenous people if there is no peace for the Colombian people. If you look at that phrase, there is a lot of content in it for us, but also the warranty of social, cultural and economic rights. In particular, peace must promote the rights of Mother Earth. People must understand that we are destroying our own life and this has to stop. That is why we speak of territorial peace. If there is no territorial peace, there can be no peace for the Indigenous people.*

Within the Nasa cosmogony, creator spirits have tasked the Nasa with looking after the world for those who are to inherit it. Mayor explained to me that, amidst the conflict and the violence, the spirits were upset and uncomfortable. They had no peace and had started sending signs such as natural disasters to warn people. So the Nasa needed to do what was necessary to establish *Wët wët fizenxi* or Living Well. Mayor explained that:

> *There must be spiritual peace. [...] We come from the creator spirits and we feel the duty to look after what they left us to inherit for the others. Peace must also be an exercise of offering, of restoring all those spiritual spaces, because they are also unsettled. They are worried too and upset. They are sending signals to tell us they are worried, what scientifically is called natural phenomena. So this is a situation that we have to stop, and it must be done through collective rituality. So peace has to be about understanding, practicing and establishing all of these, so that we can actually attain* Wët wët fizenxi *and live nicely, happily in this territory that we call Colombia.*

Mining was seen as an obstacle to Living Well because it carried the disharmony and risks to Indigenous rights that I examined earlier. As a result, most believed stopping mining was a way to build peace, but also had a conviction that simply stopping it would not suffice. Agata, for example, described peacebuilding in Canoas with examples of community actions to bring mining to a halt. Alegria said that building peace was ending mining in the Resguardo. Several others stressed that the priority for the Cabildo should be asserting the mandate of a NO to mining, because otherwise there could be no peace. However, in seeking responses for this conflict, the participants were acutely aware that more livelihood alternatives were necessary, in particular for young people and former FARC combatants. Furthermore, the participants saw spiritual work as a parallel and compulsory course of action to respond to the disharmony within the community.

Beyond what Canoas did about mining, there were other matters, not fully within the control of the Nasa, that were crucial to the path that mining, conflict and peace dynamics would follow in Canoas. Paez, Mayor and others were very emphatic that a Peace Agreement was not peace, but an opportunity to begin the reconciliation. However, they saw limited political will to engage with the requirements of reconciling the Colombian society. A core requirement for the Nasa was addressing land distribution problems they saw as underlying economic hardship. Paez, for example, feared that new armed groups were forming and that individuals would continue to prioritise themselves above their communities. In the meantime, he felt that the needs of the victims were disregarded, while questions about former combatants and FARC dissidences were left unanswered with obvious consequences for the possible expansion of illegal mining in Indigenous territories.

Canoas Responses to Mining

The Canoas participants reflected both on what Canoas had done to respond to mining and on what remained to be done. In the following pages, I briefly discuss some examples of actions the community had implemented. Later, I concentrate on some of the actions and shifts the participants considered crucial going forward.

Examples of Implemented Responses: "Blocking and Unblocking Tunnels"

The Canoas community maintained an active stance against mining. However, the process had not been without its setbacks. Spiritual work, political discourse emphasising a strong intent for decisive enforcement of mandates, children's education, extensive and costly Guardia territorial control, water quality monitoring, and environmental education and livelihood projects were examples. What several participants saw as setbacks included the election of people with conflicts of interest to the Gobernador(a) role, an inability to think about economic dynamics in the long-term and lack of funding. Some believed these difficulties had hindered effective responses. In the following paragraphs, I briefly outline some of the responses to illustrate their scale and diversity. Much of what ACIN had implemented could be seen in Canoas, or Canoas pioneered it, so I shall avoid unnecessary repetition.

Views on how effective the Cabildos had been in responding to mining varied widely. Some believed the Cabildos had lacked economic foresight to respond to the lack of livelihood options for the younger generations. Others saw the dearth of funding as the core reason for an apparent inability to sufficiently prioritise miners' livelihood projects; in addition to the political difficulties that result if miners receive higher priority than agriculturalists, working within the law, with much lower financial gain. Nasawala thought more complex political difficulties emerged when miners managed to elect one of their own, or a relative, to the Gobernador(a) role. Such lack of independence of the Indigenous Authority from the miners had, to several participants, made at least one previous Cabildo unable to assert the NO to mining mandate.

Despite those difficulties, the Canoas Resguardo, its Cabildo and Guardia had implemented extensive efforts to respond to gold mining proposals and illegal gold mining. Furthermore, the interviews I conducted during early 2017 pointed to decisive action against mining during the rest of that year. The violence had led the Cabildo to reinvigorate actions to stop mining and to test new tactics. The Cabildo that was beginning its mandate during my stay in Canoas were determined to implement territorial control to enforce observance of the community mandate not to practice mining. They were interested in understanding the miners' rationale and needs. Although members of previous Cabildos were clear on the miners' positions, the one year, simultaneous, terms of Cabildo members impaired organisational memory. So dialogue was necessary. More so given the concerns of Cabildo and Guardia about the community costs of controlling mining, and particularly considering that Government institutions that would appear natural allies were not always. Enrique (pseudonym) explained to me that in 2017, the Cabildo was going to collaborate with the Regional Attorney when facing illegal miners, but was apprehensive about interactions with the Regional Natural Resource Management Office, who had failed to consult the community before issuing licences to miners.

As mining appealed to young people and considering the fundamental role Nasa moral values have in informing choices regarding livelihoods, children's education was central to Canoas's response to mining. La Profe believed through education the community could encourage younger generations to maintain a strong attachment to Mother Earth. Environmental education was in the local primary and secondary school curricula. When I enquired about the response of children from mining families to environmental education, La Profe explained the children were mostly on-board, just embarrassed that "*the parents don't listen*" and in some cases a bit torn that "*if he doesn't go then he cannot buy me clothes*". She thought it was important for the children to question themselves that way.

The response to existing mines was different and came in the form of territorial control (or enforcement) actions. The community had identified water impacts from the mines that had to be dealt with. Significant energy and resources were put into nipping mining in the bud and into closing established tunnels. The Guardia exercise territorial control, collect information to maintain the Authorities informed about the dynamics of mining and enforce the decisions of the community. Luz Dary told me about the efforts of the Guardia. She was concerned that the costs of territorial control and corrective action in response to mining were too high for the community. Referring to the cost of responding to the Munchique violence, Luz Dary told me:

> *Because of the results of an assembly where the decision was to stop mining [...] they said that the miners who were not allowed to work in Canoas should not show up over there [in Munchique]. The mistake was that they signed an agreement that, if our people went there and were caught, our Resguardo would take responsibility. It has been very difficult to watch over [the Canoas miners captured in Munchique]. Right now there are four Guardia and four Cabildo members watching over those boys. It costs a lot. We must send them food. We must transport them. We must do shifts. The Guardia is voluntary and has no resources.*

Nevertheless, the participants thought the Guardia and territorial control work were effective responses to the problem of mining, particularly so when done in collaboration with Afro-descendant and campesino communities. Paez recounted that around 2013, they "*did a big minga [or working bee] because that whole mountain over there, they were making mining tunnels in it. So straight away we did a big minga with the Afros, all of us. The Afros are most astute. All the waterholes they have down in the lowlands of the Afros, that is water that comes from up here in the Cerro Munchique. So they sent us a lot of personnel and we went and blocked all of that and we moved that step forward. That had a big impact on our territory. ACIN came with the Councillors. It was a big minga*".

Complementing territorial control, the Cabildo and ACIN had implemented some research and livelihood projects. There was a research project from ACIN's Casa de Pensamiento that covered Canoas and examined the mining dynamics there. In 2017, the Cabildo with the One Earth Future Foundation conducted a participatory water monitoring project to examine the status of water in the Resguardo, in the context of a range of economic activities including mining. Members of the Cabildo had told me such a project was a priority for their mandate because it would provide much needed information to enrich the debate on mining and make it about what was happening rather than only about positions. Livelihood assistance for the miners consisted of Tul farms or traditional veggie patch start-up inputs, to encourage miners to establish cash crops such as coffee, and subsistence crops. However, the general perception was that the miners did not work the land, and used lack of livelihood options only as an excuse.

A Way Forward: "May Man Forget there is Gold in the Earth"

Most participants from Canoas thought the way forward was to end mining in their territory, taking a number of steps in the domains of law and rights, community capacity, dialogue, livelihood diversification and peacebuilding. In the area of law and rights, rights to the territory and to Living Well were seen as the platform to respond to external mining proposals or unauthorised entry. The participants believed it was necessary to reassert the NO to mining mandate by, for example, continuing collaborative territorial control. When it came to mining by Nasa people, the participants hoped to see further developments in Indigenous legislation. Canoas Indigenous law experts were studying avenues for a judiciary response to unauthorised mining. They were examining the concept of 'crime against nature' within Indigenous law and culturally appropriate penalties. The ordinary judiciary system and the Special Indigenous Jurisdiction overlapped in this case: unauthorised mining was also illegal in the mainstream system. The debate and the tension were about how to negotiate the overlap while maintaining autonomy. Detractors and defenders of mining utilised clashing interpretations of autonomy in this context. According to Pil, the miners argued that calling the Regional Attorney would show weakness; while numerous Nasa from Canoas denounced the illegal miners at the Regional Attorney office and asked the Cabildo to let the ordinary judiciary act and capture the illegal miners. The steps to build community capacity to face mining and favour other livelihood alternatives included actions to restore spiritual balance, in particular in young people, but also political and environmental education to promote commitment with the community and with custodianship of the territory. Many of the participants highlighted the need for the Cabildo to prioritise dialogue with the miners of the Resguardo so that they could be heard. To some, like Alegria, this should be a top Cabildo priority.

On the question of livelihoods, it seemed the community needed more information to identify the best course of action. It was clear that agricultural livelihoods were not sufficient to provide employment for all the young people in the Resguardo. Paez and Pedro had seen the success of investment in agricultural projects for miners and believed that with more support to extend the coverage, other miners could be persuaded to return to agriculture. Repeatedly, people explained that was what the miners said they need. Ul and Paez said they understood why, in the absence of Cabildo assistance, some miners continue with the activity. However, to other participants such as Luz Dary and Ortensia, it was unclear whether such assistance was effective in motivating the shift away from mining. Nevertheless, there was consensus that the community would need to work to create new livelihood alternatives and it needed significant external support.

Finally, there was an important component of peacebuilding work ahead for the Nasa of Canoas because all the steps already mentioned were contingent on establishing territorial peace. According to Mayor, the Nasa had to continue to remind all parties that "*the peace agreement has not arrived in the territory*" and the Nasa needed to see concrete action. On the crucial question of ex-combatant livelihoods, Mayor explained, "*We have told FARC: if there are militants, or combatants, or commanders, or squadrons, or dissidences involved in mining, you must remove that situation. Because the big mistake that the guerrillas made was to arm their militias in our territory. So nowadays those weapons are here*". Clearly, aside from the livelihood projects the community might be able to offer, it was mostly the responsibility of the parties to the Peace Agreement to ensure that former combatants did not become part of the illicit mining economy. Concerned about the risk of FARC ex-combatants resorting to mining, Pil stressed the unfulfilled Government responsibility to define and implement policies for social investment and reintegration. In their absence, he assured me, the problems with illicit economies like mining would continue. Furthermore, Mayor said to me, "*We have told the Government: you must control mining and control the influx of money. We know 80 per cent [of the gold] leaves [the country] illegally. You must exercise control over there. Let us control within our territory. We will propose alternatives to mining. We are asking you to persecute the illegal actors, in this case the paramilitaries that are trying to coopt mining and drug trafficking*". It was clear that the work towards territorial peace had to articulate mainstream and Indigenous Authorities.

The participants saw the spiritual domain as the start and end of the road ahead. Most Canoas participants agreed that once the Cabildo had done everything in its power to enforce community mandates, the Traditional Doctors and the community would continue with their spiritual work so that mining would no longer have a place in the minds of people and there could be peace within the community. Guia told me:

> *We've sat down, a group of Elders, to look into the future, because there will be confrontations between the two Resguardos. The families are already looking for retribution. An internal war is more dangerous than any other. In the Cabildo, we are looking at how we will work, spiritually speaking, the Traditional Authority and the Ancestral Authority, so that we do not fuel the fire. The advantage is that now the Northern Zone will join us. After that, traditional medicine has to continue so that man may forget that there is gold in the Earth.*

Discussion

The tensions, conflicts and difficulties the Canoas community faced with mining reflected long-standing debates about what constituted an appropriate relationship with the natural world and a sound way of making a living, contests about access to land within and without the community, conditions of material scarcity, the dynamics of violence and armed conflict, and the complexities of building territorial peace. In Canoas, we find the case of a community that banned mining and saw gold as a powerful mineral to be treated with respect, facing the persistent commitment to unsanctioned gold mining by some of its own community members. Like with previous waves of illicit economic activity, mining had driven fissures within the community, caused social and environmental deterioration and driven periods of intensifying violence. It should not come as a surprise that many in the Canoas community were worried about how mining dynamics would interact with the social, economic and political shifts the North Cauca Zone was to face with the arrival of former FARC combatants, in search of a way to sustain themselves.

The view that gold itself affects people's attitudes and behaviour is an example of the way that Nasa people conceptualise the natural world as having powers to affect humans and their social actions, and to apply corrective measures. The notion of 'harmony' is in some ways a reference to the reciprocal relationship that humans have with their environment. Just as misuse of nature upsets an ideal balance, so the imbalance in the environment impinges on people's lives. Gold appears to have particular powers that can evoke human desires for wealth that are disruptive, antisocial and immoral. Given that former traditional uses required that extraction and use of gold be moderated and managed by ritual experts, it seems that gold has the capacity to 'punish' people who take too much. The moral propensities of the natural world, whether thought of as spirits or simply supernatural forces, are effectively 'actors' in the Nasa moral universe. As the actions of the miners deviate from the notion of harmony or the ideal life described by *Wët wët fizenxi*, the broader community fears that Mother Earth might act decisively to restore the natural and moral balance that they have failed to maintain.

As long as crucial questions about livelihood options remain unanswered, mining will continue to be an economic alternative for those in need, but also a platform for new ambitions and for political contests within the community as well as from external agents. It is clear that in Canoas the question of alternative livelihoods for miners was yet to be fully explored and was plagued with internal political conundrums. Furthermore, it was well acknowledged that saturation of the local agricultural markets was a difficulty for Canoas agriculturalists. The miners of Canoas had already 'touched gold' and this raised the question as to whether traditional agriculture could provide sufficient incomes in the micro-economic context of the post-agreement transition in North Cauca.

While the community effort to respond to mining was extensive, there were still areas that the Canoas inhabitants could further explore, although it was difficult for the Nasa community to respond on its own to the violent and illegal groups connected to some mining economies. There were spaces where the Canoas community could still have latitude to act internally. For example, considering the family impacts of mining moneys and the humiliation routines male miners are said to engage in, an honest examination of the types of masculinity that young people favour, and the community and external reinforcing factors, was in order. A dialogue on this vital social trend was necessary and was consistent with the Origin Law principles that call on the Nasa to maintain dialogue with each other. But the people of Canoas were aware that they needed to collaborate with others. They had identified allies in Afro-descendant and campesino communities, were exploring new ways of collaborating with some National State agencies and expected the State to assume its responsibility to stop illicit mining network expansion.

The experiences and feelings of Canoas participants about mining reflect the possibilities and difficulties of what Boege (2010) has referred to as post-conflict "*hybrid political orders*". These are political orders where various providers of security and justice coexist in an environment of mutual and positive accommodation. Boege argues that post-conflict peacebuilding responses that acknowledge this hybridity offer a far more promising path than those that centre on the notion of a Weberian state. The Nasa organisations are far more than justice and security providers. They provide environmental, agricultural production, educational and health services, and they are producers and enforcers of moral content. They can approach the problems with mining and violence with a range of culturally and territorially specific resources more appropriate than direct State force intervention. But in responding to mining, to the armed conflict and to their intersections, the Nasa often found themselves lacking sufficient support or complementary action from the mainstream State authorities. This placed Nasa organisations and Authorities on the frontline of the confrontation with dangerous violent and criminal actors. Where the State lacked the capacity or the political will to engage actively in the hybridity required for territorial peace, the customary actors of the hybrid political order carried an excessive burden and human cost.

Note

1 Understood as both negative peace, which refers to the end of violence and conflict, and positive peace, which relates to attaining equity, cooperation and harmony (Galtung, 2015, p. 622).

References

ACIN. (2011). *Plan territorial cultural*. Santander de Quilichao, Colombia: ACIN.

ACIN. (2013). *Cabildo de Canoas*. Retrieved January, 2018, from http://anterior.nasaacin.org/index.php/planes-de-vida-2013/plan-de-vida-yu-luucx/cabildo-de-canoas.

ACIN. (n.d.). *Plan de Vida Yu'Lucx – Resguardos de Munchique Los Tigres y Canoas*. Retrieved January, 2018, from https://nasaacin.org/plan-de-vida-yu-luucx/.

Boege, V. (2010). How to maintain peace and security in a post-conflict hybrid political order: The case of Bougainville. *Journal of International Peacekeeping*, 14(3–4), 330–352. doi:10.1163/187541110X504382.

Canoas Indigenous Reserve Environmental Rules. (2015). *Reglamento Ambiental del Resguardo Indígena de Canoas*. Canoas, Colombia. Canoas Cabildo.

Canoas Cabildo (Canoas). (2017). *Identificación de los impactos de la actividad minera en el territorio Indígena Resguardo de Canoas - propuesta*. Santander de Quilichao, Colombia: Canoas Cabildo.

Caro Galvis, C. & Valencia, Y. (2012). El caso de pequeñas y medianas minerías en el Cauca: ¿Alternativas o amenazas a la autonomía indígena?. *Señas*, 2, 17–27.

Galtung, J. (2015). *International encyclopedia of the social & behavioral sciences.* Waltham, MA: Elsevier Ltd.

7 Violence, Peace and Mining

Introduction

The behaviour and roles of local actors in post-conflict peacebuilding and post-conflict mineral resource governance have been important concerns in the academic literature. Most of this literature has adopted an outsider's perspective. At the centre are international actors' actions. Localised actions are, more often than not, construed as responses to what international peacebuilding actors do. As a result, this literature has tended to prescribe roles for, or ways to influence, local actors. However, in the last decade, the peacebuilding literature has given more attention to the 'everyday' nature of peacebuilding and its deeply rooted local dynamics. Similarly, the literature on post-conflict high value natural resource governance, having attempted to prescribe roles for local actors, or left them mostly unacknowledged (Chapagain & Sanio, 2012; Lujala & Rustad, 2012), now seeks to better understand their behaviour. Having examined the Nasa response to mining, in this chapter I explore its links to Nasa peacebuilding work, recognising the proactive role of the Nasa in these processes.

The Nasa of North Cauca have a complex response to mining in their territory and a tradition of peacemaking and peacebuilding work. These contrast with what the Nasa perceive as lack of action from traditional peacebuilding actors in regards to mining. The premise that the locals, in this case the Nasa, merely respond to, or resist, what international actors do does not hold in this context. Rather than trying to force the Nasa into a one-size-fits-all notion of post-conflict mineral resource governance, the task is to understand how they approach it in light of their priorities, institutions, values and perspectives. This is what I set out to do here.

Participants understood that the link between responses to mining and building peace integrated: notions of peace and harmony, the moral economic framework, livelihood considerations and experiences or memories of mining. Either resisting or responding to mining could constitute peacebuilding. Mining affected livelihoods, collective and individual rights, and community harmony. It created moral tensions. Participants' experiences of mining and mining conflicts informed whether they saw resisting mining or responding to

DOI: 10.4324/9781003226895-7

protect mining livelihoods as forms of peacebuilding. However, neither livelihoods nor moral principles fully determined how people thought mining was to be managed to build peace. What people believed would bring harmony to their communities was a consideration as well. It would be simplistic to say that all Indigenous miners believed building peace meant allowing the Indigenous miners to continue. It would also be simplistic to assume that all who opposed mining believed it could not coexist with peace in Indigenous territories. Understandings of peace and of what would bring community harmony were as fundamental in informing people's views on mining as were livelihood and moral principles.

In this chapter, I concentrate on Nasa ideas of peace and activities to achieve peace, to understand their link to Nasa responses to mining. I begin by discussing the limitations of prescriptive, subordinative or binary (local vs. global) perspectives on the roles of local actors in peacebuilding – a critique that equally applies to post-conflict mineral resource governance. Later, I examine how the participants understood the relationship between mining, violence and peace. A brief discussion of the Nasa resistance identity and process will follow. These will foreground an analysis of Nasa ideas on peace and peacemaking and peacebuilding work. Based on these, I explore the links between the Nasa responses to mining (as outlined in Chapters 5 and 6) and peacebuilding work.

A central element in this chapter is the armed conflict, and there are no better narrators of it than the communities who experience it directly. I shall frequently and extensively refer to statements from the Nasa participants. Their explanations and narratives, even in translation, should lend tangibility to the lived experiences that have shaped Nasa peacebuilding identity and practices. In doing this, I seek to provide the reader better access to the Nasa way of thinking about the Colombian conflict and about peace. As González Rosas (2016) has argued, using direct testimonies allows us to illustrate contrasting perspectives within the Indigenous movement to exemplify the tensions. Furthermore, by presenting these statements, I seek to pay respect to community resistance to the armed conflict.

From Passive Recipients to Globally Networked Local Agencies

To understand the relationship between Nasa responses to mining and peacebuilding work, it is necessary to consider the role of the Nasa as local peacebuilders. The academic literature on peacebuilding has examined the roles of local actors in what some refer to as the "*local turn*" (Mac Ginty, 2015, p. 841; Mac Ginty & Richmond, 2013) or the "*move towards the 'local'*" (Richmond, 2010, p. 671), a renewed interest in the idea of the local as a source of solutions (Mac Ginty & Richmond, 2013). This literature sought to explain local-international actors' interaction in the context of post-conflict peacebuilding. In doing so, it has largely taken the standpoint of peacebuilding actors who do not see themselves as localised, and has analysed and made recommendations on how those actors might understand and approach the local (Jabri, 2013; Mac Ginty, 2008; Richmond, 2010). This literature has advocated that 'internationals' should embrace local capacities and see local people as a 'resource' to influence and enable, rather than as a hindrance (Lederach, 1995). It has described the tension between globalised peacebuilding and local agencies and resistance, and how hybrid peace emerges from these tensions (Mac Ginty, 2010; Richmond, 2010, 2015). Bringing attention to autonomous agencies and acknowledging their need for self-determination, it has highlighted the role of the 'everyday' (Richmond, 2010) and has recognised that attention must shift to "*practices, ideas and cognitive frameworks*" behind responses to post-conflict transformations (Chandler, 2013, p. 25). This literature has stressed that the "*everyday*" is the realm of inductive and critical approaches rather than of formal explanation (Richmond, 2010, p. 683). The latter point signals a move beyond what I see as the core limitation of political and international relations approaches to the question of local peacebuilding efforts: the emphasis in understanding local communities' role from outside that construes them as passive or at best as resisting or only responding on the basis of what external actors have done (Mac Ginty, 2010). What the peacebuilding literature refers to as the everyday, i.e. the quotidian acts and processes of peacebuilding, belongs in the realm of ethnographic inquiry, where this study is located.

In analysing past and future Nasa responses to mining, I shall avoid portraying them as merely responses to what traditional peacebuilding actors do. Considering the extensive efforts of Nasa Authorities to govern mining dynamics and their multiple activities in the field of peace, the model of passivity or resistance is wanting. Besides, local Indigenous organisations in North Cauca perceived that traditional peacebuilding actors had either left local mining dynamics unaddressed or had limited scope for influence. When it came to State actors, the widely held perception amongst research participants was that, through corrupt practices, members of the State apparatus were linked to the criminal organisations behind violent forms of mining. In this "*hybrid political order*"[1] (Boege, 2010, p. 1) the Nasa, as the customary actors, were more active and more present in the territory than State actors and carried an excessive burden (see Chapter 5). Crucially, the Peace Agreement with FARC did not address mining directly. It set out principles for sustainability and inclusion, and laid out mechanisms for participation that could be used to respond to mining dynamics. However, mining was an established illegal economy, central to the finance portfolios of illegal armed groups (Rettberg & Ortiz-Riomalo, 2016). This suggests that mining warranted a direct response within the Agreement. In other respects, interviews for this research revealed that by late 2016, only months before the Agreement was finalised, and ahead of the Peace Agreement plebiscite, key State mining institutions were yet to engage in analysis and planning on how to address mining dynamics in the post-conflict period. Some of them understood the link between mining and violence as strictly a law and order matter.

Even if this were the case, interviews with regional State actors revealed ongoing concerns about links between illegal miners and members of the armed forces. In sum, participants described a situation where a State response to mining within the context of conflict-to-peace transitions was far from consolidated and, in some specific cases, the actions of State actors were counterproductive.

Nasa responses to mining and Nasa peacebuilding occur in a context where local actors participate in globalised processes characterised by friction (Tsing, 2005) between ideas and economic processes, none of which is merely 'local' or 'global', liberal or illiberal.[2] The Nasa are active members of transnational activist networks working on Indigenous rights, human rights and the environment. Here the global loses visibility because what is more identifiable is its local expression, but it is still global. International peacebuilding actors are not the primary force, nor do the locals monopolise. The efforts of the Nasa in relation to mining and the armed conflict are not mere resistance to external global processes, neither are they 'pure' local solutions, free of exclusion and discrimination, or devoid of globalised elements.

What follows has a distinct local character but it also reflects intricately linked local and globalised processes. Behind the steady advance of the 'yellow machine' through the rivers and the mountains where the Nasa lead their lives are the global price of gold and the workings of its international market, capitalist pursuits run from afar, desperate measures to feed a family, rent seeking behaviour or calculated risk taking, and entrepreneurship. Illegal armed groups represent the local, violent expressions of global ideological and economic contests, including disparate views about development, progress, economic opportunity and a fair society, and disputes for control over natural resources.

Violence and Mining

> *First, the paramilitaries arrived here, in [19]91, that was the first massacre, at the El Nilo farm. They started gathering force in the territory and started to get organised and in 2000, started to kill in such a way. They did more massacres. There were several massacres like the Gualanday massacre, the San Pedro massacre, the Alto Naya massacre. There were several massacres to sow terror in North Cauca. There were targeted murders: an Indigenous person would leave Santander and their body would be thrown in the Cauca River at La Valsa.*
>
> [...]
>
> *Then the paramilitaries 'demobilised' and then a new situation ensued with the guerrillas who began to massacre people whom they labelled as snitches, informants, paramilitaries. So, FARC murdered people, but so did the paramilitaries.*
>
> (La Mariposa (pseudonym), Nasa leader)

Historically, Colombia's successive violent political contests left Indigenous people dispossessed and stigmatised, so the Nasa mobilise resistance against activities linked to violence. Violence has been a force shaping Colombian politics (Escobar, 1995; Rappaport, 2000; Taussig, 1984). During the nineteenth century numerous contentions led to a reiterative war pattern that has persisted. The disputes were about the government model, revealing tensions between centralist and federalist government; about Colombia's entrance into the world economic order, and the resulting modernising trends; as well as about the role of the subordinate classes in political life and the role of the church (González González, 2014). According to González González (2014), these wars set patterns in the ruling parties: one of reluctance towards social mobilisation in the Liberal party and one of 'fear of the populace' amongst the Conservatives. Resulting from the violence, argues González González (2014), came partisan identities used to make sense of violent experiences. As a contemporary reflection of that fear of the populace, the Nasa are heavily stigmatised as a violent people, their role relegated to resistance, devaluing their participation in the construction of the Colombian republic (Rappaport, 2000). Rappaport stated that violence has been "*a constant reality in the history of the Nasa*" and had turned into a force that exerted control over Nasa communities (Rappaport, 2000, p. 145). Despite and because of these, the stance of the Nasa organisations is to resist that violence in their territories and the economic activities that are contingent on violence.

Views on the mining-violence-peace relationship in Indigenous North Cauca were diverse, but one can discern a prevalent view. A majority did not practice mining and saw it as an obstacle to peace. In the eyes of many in the movement and its communities, mining breeds internal divisions, conflicts with external parties and the violence and armed actor presence that usually accompany illicit economic activity. Nevertheless, there was contestation because some Nasa individuals and Authorities were involved in mining (see Chapter 4). The divisions around mining are harder to face than conflicts with outsiders and have escalated into confrontations and even intra-community violence.

Conflict with outsiders may come in the form of tensions with formal mining companies seeking entry into the Nasa territory or conflict with illegal mining groups. The latter has at times escalated to violence. However, the distinction between illegal and legal mining is not always clear. There are perceptions in the communities that there are connections between legal and illegal mining. In 2015, ACIN's studies concluded that a large proportion of Nasa territory was titled to mining companies without consultation – effectively, a threat to Indigenous autonomy. Several participants blamed mining companies for the arrival of illegal mining. In their eyes, company presence sent a signal that there was gold in Indigenous territories. Some others believed that formal mining companies would follow once illegal miners had entered Indigenous territories. Evaluating the veracity of these claims is not within the scope of this research. However, the key fact remains that people perceive links between the formal, large-scale mining companies, and illegal miners. The Cauca Regional Attorney's research has found cases of companies that articulate illegal and legal activity in Cauca. Termed 'amphibious' mining companies, these are legal gold mining companies that source illegally mined gold from Cauca and commercialise it through the regular and legal channels of front companies in Medellin, attributing the production to legal mines. The link with illegality draws yet another connection with violence. Thus, several Indigenous participants considered that a range of mining activities were connected with illegality and violence.

In connection to mining, participants cited the presence of illegal armed groups, threats against vocal leaders or community members, and direct acts of violence. The Cauca Regional Attorney reported that in 2016 and up to March 2017, three North Cauca leaders had been murdered because of their opposition to mining. The trend had continued, according to Colombia's Ombudsman statements in 2018 (Cárdenas, 2018). Furthermore, members of the Indigenous movement have a clear understanding that illegal mining contributes to financing armed actor presence.

An often-cited anecdote about illegal mining and violence illustrates why people see illegal mining as an obstacle to peace in Indigenous territories. A number of participants narrated an incident in the Canoas Resguardo where the community gathered in large numbers to expel a group of miners who had been asked to leave several times. According to participants, the community approached the miners and the miners responded by firing shots into the air or towards the community. The community continued approaching and set an excavator machine on fire. The miners then fled. Participants used this story to explain that illegal miners do not respect Indigenous territories or Indigenous lives, that they respond to Indigenous opposition with negligence or violence, and are only driven away through the use of force. There was also much suspicion and distrust among Indigenous participants about mining companies in general. In this sensitive and polarised context, it would have been difficult to include here company perspectives about the situation in Cauca without being seen as taking sides and possibly exacerbating negative perceptions or tensions. This was an unnecessary risk because explaining mining company behaviour was not an objective of the study. For a brief discussion of examples of industry perspectives, please refer to Chapter 8.

Through prolonged experience of the dynamics of violence, illegal economies and armed conflict, the Nasa have developed a profound understanding and apprehension that permeate their views of other illegal economies such as mining. Parallels between the drug economy and mining were conspicuous. Having lived side by side with it for decades, the Nasa know what happens to communities with drug production and trafficking. They understand drug trafficking as contingent upon, and supportive of, the violence of illegal armed groups. Community members enumerate the signs of the drug trade like the symptoms of an illness – the 'sumptuary consumption', the decline of traditional agriculture, substance abuse, domestic violence, unexplained deaths and suicides, murders, violent confrontations and threats. They have identified those same symptoms when illegal mining has come to the

Resguardos. There were several anecdotes about this. For example, Sek (pseudonym) told me about Canoas:

> *Things started changing in Canoas in 2010 and 2011. People who cultivated fruit started to just do mining [...] in 2011 there was a suicide, a young person turned up hung at home, he was a miner. So we started to link family deterioration with mining. So what started to happen was: if I had more money than you, then I could take your woman from you. And the clothes of this young person they disappeared. It is a strange thing and his woman said that he worked 20 days in the tunnel. But we didn't find gold. That was a mystery.*
>
> [...]
>
> *Since 2004, with the demobilisation of paramilitaries, there was calm in Canoas. You wouldn't see armed groups here. But from the arrival of mining we also saw armed groups. We don't know who they were, they didn't identify themselves. There was a night when 15 or 20 armed men with baklavas turned up. They were looking for some of the people who had mining mills.*

Sek and others saw striking similarities between such incidents and what they witnessed with the drug trade. Knowing where drug production and trafficking have led, the participants could not but draw parallels and arrive at similar conclusions. Michael Taussig (2004) has written extensively about the ties between gold and cocaine as commodities and as substances with close connections with evil (and good) and with violence. To the Nasa, mining, like drug production and trafficking, brings violence, gives hold to illegal armed groups and deteriorates the social fabric of their communities.

Where communities seek to move away from the violent dynamics of the drug trade, mining appears to facilitate them and build on existing infrastructures of violence, both those that apply it and enable it, and those that allow it to modify behaviours and ways of thinking. Taussig (1984), speaking about terror, explained how important the latter are:

> *[...] the victimizer needs the victim for the purpose of making truth, objectifying the victimizer's fantasies in the discourse of the other. To be sure, the torturer's desire is also prosaic: to acquire information, to act in concert with large-scale economic strategies elaborated by the masters and exigencies of production. Yet equally if not more important is the need to control massive populations through the cultural elaboration of fear.*
>
> (p. 469)

The violence infrastructure that emerges enables violence to circulate. It facilitates what the violent seeks, through direct acts of violence and through the "*cultural elaboration of fear*" that ultimately changes behaviours and ways of understanding amongst the victims. Where people might aspire to the vision of a utopian society – the idea of *Wët wët fizenxi* with reciprocity and embeddedness at the core – illegal economies linked to the armed groups pull communities into a dystopian sphere of capitalist pursuit, where people not only take without giving back to the community, but also mediate relationships through violence. The violence is both brutal and born of greed.

Nasa Resistance Identity

Nasa communities and organisations have a strong resistance identity. The Nasa, in their discourse, identify themselves as a resistance peacebuilder movement that protects human and collective rights, life and the territory in the context of historic and ongoing violent, systematic dispossession, oppression and colonisation. Nasa morality attributes high value to connection with Nasa historical resistance and resistance is embedded in the Nasa habitus[3] (Rappaport, 2000). This is manifest in the ways the Nasa organise themselves and operate. As La Mariposa, a Nasa leader, explained, resistance has contributed to shape the Nasa organisations and the way they work. I heard and read the word resistance repeatedly throughout my days with the Nasa and when away. When calling my collaborators from Australia, I would ask: "*How are you going? What is happening over there?*" and the reply was always something along the lines of, "*We are well, things are difficult, but we continue to resist*", or "*we face lots of problems, but we are here, resisting*". Frequently formal letters or emails would close with "*saludos de resistencia*" which roughly translates as "*we salute you, in resistance*". Resistance is a constant reference and a keyword that positions the Nasa speaker in relation to listeners. It is with pride that the Nasa invoke their resistance identity. For example, Julián (pseudonym) told me:

> *I feel proud being part of a resistance struggle that goes back to the arrival of the Spanish in 1492. Like the struggle of Cacica La Gaitana, a very courageous one; the fight of Juan Tama, of Quintín Lame; the struggle of our compañeros who sacrificed even their lives back then in the year 1971 in the land recovery campaigns; and the fight we've had to give after the 1991 Constitution to exercise our autonomies.*

Nasa people understand much of their daily existence and their history through reference to resistance tactics, which are linked to their cosmogony (Rappaport, 2000, p.17). They have been part of the oppressed rural working class. For Scott (1985, p. 290), class resistance comprises "*any act(s) by member(s) of a subordinate class that is or are intended either to mitigate or deny claims […] made on that class by superordinate classes […] or to advance its own claims […] vis-a-vis those superordinate classes*". Before the formation of the Republic, Nasa resistance tactics consolidated in the form of prominent leaders or Caciques, significant battles, legalisation of Indigenous territories and territorialised commemoratory practices – such as walking the

Resguardo limits; in all of these, legalistic resistance played a significant role (Rappaport, 2000). There is continuity to these practices (González Rosas, 2016). This is evident in the prominent role of individual leaders, in the territorialised resistance and in the legalistic focus of the Nasa struggle. A different kind of 'battle' occurs now, but victories against a violent aggressor remain a motive of commemoration and celebration.

With mining and the post-conflict transition in mind, it is significant that much of the resistance takes purpose and meaning in opposition to State expansion projects and to the armed conflict (González Rosas, 2016; Rappaport, 2000). A 'battle' might be standing up to armed illegal miners or criminal bands or expelling drug processing facilities from an Indigenous territory. It might be declaring and enforcing that armed groups, legal or not, will have no access to an area because it is a 'territory of peace'. It might be a mass mobilisation against the armed conflict. La Mariposa once told me that after so many massacres:

> *[…] all that led us to embark on the first mass mobilisation, in 2001, from Santander de Quilichao to Cali. For the first time in history, we walked from here to Cali. It was very hard for us because we are used to walking in the mountains, but to have to walk on the Pan-American highway that is flat, that was horrible for us, it gave us blisters. We weren't used to that, the pavement was very hard, but we resisted and arrived in Cali. We told them that we were an Indigenous people, that if they were going to massacre us and kill us, then we would make a lot of noise in the streets of the cities, wherever, but we would not let ourselves be killed.*

The resistance is strongly linked to the territory, its legitimacy and integrity. The Nasa regularly commemorate the legal victories in achieving recognition of Resguardo territories. Rappaport (2000) stated that the resistance tradition of the Nasa has profound roots in the past and the sacred landscape provides a space for past and present resistance to come together. That landscape and its sacred qualities are in danger because of violence and primitive accumulation[4] of land, land grabbing (Cramer & Wood, 2017; Grajales, 2011), armed conflict, and some types of mining. The Nasa resistance often seeks to assert the legitimacy of Indigenous territory (Rappaport, 2000). Resistance against violence, armed conflict, mining and State expansion seeks to preserve territorial integrity.

Frequently, the Nasa participants explained Nasa responses to mining and to the armed conflict based on the principle of peaceful resistance – a resistance that is both a way of doing things and an identity that informs responses. Peaceful resistance or non-violent action has gathered significant momentum as a force against corruption, oppression, human rights violations and authoritarianism in Latin America (Pugh, 2018). According to Pugh (2018), such non-violent action requires significant planning efforts, enforcing a discipline of non-violence and ongoing training as part of international networks. Consistently, the Guardia, who at times become the entire community and are one of the most visible symbols of Nasa resistance, receive training and support to strengthen their capabilities.

There are strong parallels and intersections between Nasa discourses on, and responses to, mining, and those addressed at the armed conflict. Responses to both are seen as assertions of territorial autonomy rights and as acts to protect life and collective rights. In a context of existing and proposed mining presence, State inability to control mining and ongoing armed conflict, the Nasa understand their responses to mining as proactive rather than subordinated to the actions of other peacebuilding actors or State mining governance institutions. Their resistance is a creative process guided by the political and economic aspirations of the Indigenous movement.

The Nasa Positive Peace: Hopes and Fears about the Post-agreement

One must examine Nasa ideas of peace to understand the relationships between Nasa peacebuilding and responses to mining. Let us begin by concentrating on Nasa notions of peace, which are diverse, but converge in being aligned with two concepts: the concept of positive peace, and the ideal of *Wët wët fizenxi*. First, Nasa ideas of peace as reflected in interviews and documentation consistently aligned with the notion of positive peace: going beyond the end of violence to address the structural causes of the conflict and realise rights, creating a culture of peace. Prominent peace scholar Johan Galtung defined the requirements of peace as the coming together of negative peace (or the end of violence and conflict) and positive peace (as equity, cooperation and harmony); he states that positive peace goes beyond the mutually assured survival of negative peace into building a peaceful coexistence with mutual rights and obligations that require equity and harmony (Galtung, 2015, p. 622). Second, the participants related peace to the aspirations embedded in *Wët wët fizenxi* (as discussed in Chapter 3). The synthesis of these two ideas is that peace is attaining *Wët wët fizenxi*, which is contingent not only on a state of non-violence but on harmony.

The prevalent view among Nasa participants was that the end of the armed conflict is a pre-condition for peace, but peace itself depends on addressing the sources of social conflict. Peace was possible only in the absence of violence, but the mere absence of violence was not peace. For example, a Nasa leader said in an ACIN gathering:

> *The end of the armed confrontation with one guerrilla group is not peace. It is a state of relative calm. But the underlying social conflict continues. Peace is not merely silencing the guns.*

Like this Nasa leader, Andrés (pseudonym), another prominent political leader, said ending the armed conflict would bring relative calm because it would mean "*the end of the recruitment, the end of the shootings, the end of the armed political pressure*", so that the pre-conditions for peace exist. In that sense, the Peace Agreement constitutes a building block towards peace.

Implicitly and explicitly, the Nasa linked their idea of peace to the notion of *Wët wët fizenxi.* Ideas of peace centred on maintaining harmony with Mother Earth and within communities, or on realising rights or attaining freedoms, all of which resonate tacitly with the idea of *Wët wët fizenxi.* Often people made the link explicit, referring to peace and *Wët wët fizenxi* as interdependent or even equivalent. Manuel (pseudonym), a seasoned Indigenous Guard, defined peace as "*protecting rights, human and collective, and protecting the territory*". Diego (pseudonym) concurred. He said that "*peace IS the territory, it is* Wët wët fizenxi". Others linked peace to elements embedded in this philosophy, such as freedom or harmony. For example, one of the Traditional Doctors I met preferred to talk about harmony in the communities and the territories rather than peace. Peace was too westernised a term for him. An experienced peace practitioner, Martha (pseudonym) concentrated on peace as freedom. She said, "*peace is walking freely in the territory, and visitors [pointing to me] coming and walking freely in the territory*". In pointing to me, Martha emphasised the idea that she wanted peace to be for all. Most people brought a pragmatic perspective into their ideas of peace by tying it to improving living standards. This was evident in conversations about hopes for the post-conflict transition. Most believed that, with the end of the armed confrontation, improved conditions for economic innovation and production would come. In all, Nasa peace relates to protecting people, territory and rights, to regaining lost freedoms, to living in harmony and to improving living standards. All of these are consistent with notions of positive peace and with the concept of *Wët wët fizenxi.* These ideas of peace refer strongly to core elements of the Nasa morality.

That alignment with positive peace and Living Well is visible within the Indigenous organisations of the region, as is an understanding of peace as the realisation of Indigenous aspirations. CRIC speaks of peace in terms of harmony and the ability to attain the objectives of Indigenous Life Plans. CRIC has stated that for Indigenous people, peace "*should be interpreted based on their cosmogony, their Life Plans, territory, self government exercise and on strengthening systems that will lead to harmony between families, communities, CRIC and Mother Earth*" (CRIC, n.d.). The Nasa, through their Association of Indigenous Authorities of North Cauca, ACIN, are part of CRIC. ACIN has its own Cultural Territorial Plan and the logic on peace draws similar links. For example, a senior ACIN leader, Juan, stated: "*in putting in practice our Life Plan, we are making peace operational*". To Juan peace related to attaining the multiple objectives in the Life Plans. These are ultimately the concrete interpretations of the *Wët wët fizenxi* ideal.

From Armed Conflict to Post-agreement Transition

Nasa hopes and fears about peace and about the post-agreement transition have formed within the context of lived experiences of the armed conflict. I found rich explanations and reflections on this in conversations with Juan, Nicolás and Samuel (pseudonyms) who held senior positions at ACIN. In the following sections, I share some of their descriptions on how conflict and peace dynamics affect communities and interact with ACIN's aspirations and work. Juan, Nicolás and Samuel told me about the effects of the armed conflict on Nasa collective aspirations and organisations, and about Nasa hopes and fears about the post-agreement transition.

Obstacles and Innovations

That some Nasa participate in illegal economies or in armed groups was to Juan one of the main obstacles for ACIN and its communities. To him, such involvement emerged in the overlap between conflict and material poverty in the geographic confluence of illegal economies. Juan's reflection was that:

> *Unfortunately, in our territory there are many actors including the armed groups, both illegal and legal. Unfortunately, we also have certain conditions that have allowed those groups to remain here. So we have the ill-termed 'crops of illicit use'. Our territories are in a strategic zone for these groups. The current conditions of the country and the economic circumstances contribute to our young people, and a lot of our people, to join those legal or illegal actors, so that they can have an income and be able to support their family. This is a reality that we cannot deny.*

The armed conflict has been a determinant, if brutal, force in the life of ACIN as an organisation. Nicolás has seen the conflict both consume and shape ACIN. It created obstacles and triggered creative changes. It was a colossal distraction: priorities had to be set aside to support the victims and respond to emergencies. In parallel, the events of the conflict led ACIN to create new programs. Those are now emblematic in the organisation, such as the Guardia. It is difficult to imagine ACIN without its influential Indigenous Guardia. This is how Nicolás explained the influence of the conflict:

> *In order to look into the future, we must remember the past. ACIN basically was born amid the conflict. It has survived all these years in the middle of the conflict. ACIN's objective was to support the Cabildos in political, social and cultural development. But, because of the conflict, ACIN ended up responding to the sequels of the conflict. 'Hey, there was a kidnapping here! Hey, there was a murder there! Now, there's been a violent threat made here!' So that forced ACIN to create several mechanisms and even two networks: Defence of Life*[5]*, and Justice and Harmony. That was to support the Authorities, the Cabildos, because some are strong and some are not as strong and need a lot of support. ACIN has tried to accomplish its objective of social, political and economic development, even administrative development, but what the conflict entailed took a lot of time away* […].

Hope has fuelled the Nasa in persevering despite these tragedies, difficulties and setbacks. Juan told me that they continue, driven by a belief that the efforts will eventually yield fruit:

> *As an organisation, amidst this situation, we still continue to fight for our rights, for the territory. It's not easy. It has come at the expense of the lives of many of our leaders and the suffering of our community. But we know that this struggle that we have been fighting, and that our Elders have fought, will sooner or later have to yield fruit. So from the organisation, we always fight for peace, for tranquillity, for a life of harmony and equilibrium, for Living Well. But all these external actors and all this external situation that come and enter into our territory mean that we cannot attain that reality that we imagine in our hopes […].*

Hopes

During its negotiation and early implementation, the Peace Agreement continued to cement hope. Nasa participants saw the post-conflict period as an historic opportunity to unite across differences along fault lines like culture, ethnicity, organisational affiliation or sexual orientation. In Juan's words:

> *This new scenario of peace is first an historic moment for our country. That is how we see it. I have been saying that. More so for those of us who have lived this war in the flesh. The country needs this so that we can stop thinking about wars and leave aside so much hatred, so much resentment and all these situations that these 50 years of war have left us; so that we can think about a multi-ethnic and multi-cultural country where all can live in harmony. So as an organisation that is how we think, and that is what we have been fighting to recognise: the fact that we are not on our own in the territory. There are other social actors in the territory who are fighting for the same and are being victimised by the same actors, be they legal or illegal armed actors in the territory. We are talking about our Afro compañeros, our mestizo compañeros, the campesinos, us as the Indigenous people, the social organisations, the LGTBI compañeros, other compañeros who have all been fighting and winning a space in the framework of a free and democratic country like the National Constitution says.*

Within that need for unity across social sectors, the interethnic and intercultural dimension – in other words the relationship with Afro-descendant and campesino communities – appears most decisive to peace. The Nasa organisations were determined to maintain harmony with Afro-descendant and campesino communities by approaching territorial conflicts with mutually beneficial solutions instead of the confrontation and competition that escalated past conflicts.

While bridging across old differences appears as a pre-condition for peace, in the minds of participants, attaining peace in the territory and nationally should also release communities to centre on their priorities. Nicolás hoped that a post-conflict scenario would allow organisational efforts to concentrate on the Cultural Territorial Plan and on organisational systems and goals. Those organisational goals have the ultimate aim of delivering improvements within communities. So Nicolás, like many others, hoped that innovation and creativity would see the Nasa economy flourish in a scenario of peace where they could collaborate, invest, combine their agricultural knowledge with technical support and initiate new ventures:

> *In an environment of peace we can get a lot done [...] We have very good people and they can devote themselves to working and get a lot done. There will be a big boom in economic matters and development projects because there are expectations in the communities. If ACIN and its Cabildos have a Life Plan, where they talk about an economy that is not linked to illicit crops, but rather a more diverse economy, with more opportunities and with respect for nature, that would create the right environment. So we would be thinking about things like ethno-tourism, food production but using natural fertilizers, that sort of thing.*

In a scenario where peace enables a stronger economic system, some expect their Nasa communities will have more resources to support and orient their members towards avoiding illicit economies or unauthorised activity such as mining. In Nicolás's words:

> *We come across situations that we have tried to manage, but it has been difficult, because they have intersected with the armed conflict. We are talking about the illicit crops, about mining itself trying to enter here and if the armed groups are there then they prevail. I am not saying that the communities would impose their will on others, rather that they would have an easier job in raising awareness and doing pedagogy with people. So with that perspective in mind, we see that the North Cauca situation, and I also believe Cauca and hopefully all of Colombia, will have a very important change [...].*

Ultimately, the hope is that the end of the armed confrontation will provide the conditions for autonomy and self-realisation, in other words to work on what the Nasa see as peace. Nicolás explained it thus:

> *So we decided that together we can continue to advance and we can make sure that what is being negotiated between the armed actors and the National Government can, to some extent, create the conditions for us to exercise our government and realise what is in our community sentiment.*

In the post-conflict or post-agreement transition, in the notion of seeing an end to the armed confrontation, Nasa participants saw many positive opportunities for their communities and organisations.

Fears and Uncertainties

Just as there were bright hopes, so there were fears and uncertainties about what might come in the post-agreement period: fear that transitional arrangements for justice and ex-combatant reintegration might lead to losses in Indigenous rights, territories and freedoms; fear that something worse might fill the vacuum of territorial control after FARC's retreat; fear that mining could grow in Indigenous territories. These uncertainties and fears relate to what the Nasa see as the social, economic, political and legislative forces that can deplete their rights and their territories.

As certainty grew on what would be included in the Peace Agreement, so did the uncertainties for Nasa organisations: What could happen with existing gains on autonomy, territories or special Indigenous jurisdiction on legislative and judiciary matters? Concerns swelled that the Peace Agreement could ultimately deplete some rights and, therefore, peace itself for Indigenous communities. Central to the Nasa was the risk of literally losing ground or of having to concede their rights for the sake of transitional arrangements. Samuel's statement provides an example of the apprehension people were feeling during the later months of Agreement negotiation:

> *These territories cannot be compromised ideologically or politically. There must be respect for the competencies of Indigenous Authorities, for the constitutional rights that have been won; there are many rights that have been won. There is concern because the idea of a new constitution is being promoted. If that happened there would be many interested in limiting the rights that Indigenous people have won. So we must be alert and we have made that clear in the conversations with the Government and also through our delegation to Havana.*

At the eleventh hour, there was some progress on this front that placated some people, but other uncertainties remained. The important inclusion of an Ethnic Chapter in the Peace Agreement brought guidance, at the principle level, on how the Peace Agreement would safeguard the existing rights of Indigenous people. However, the NO victory in the Peace Agreement plebiscite rang alarm bells about cross-party commitment to the Peace Agreement, in particular as regards to transitional justice balance. Furthermore, despite the differential gender impacts of the conflict, the plebiscite results required changes that watered down the gender approach of the original Agreement. The list of unresolved concerns was lengthy, but let us concentrate on those around mining.

With the post-agreement in mind, there were three main concerns about mining: i) the removal of barriers to entry in Indigenous territories for corporate miners, ii) intensifying illegal mining in the wake of FARC's retreat, and iii) the potential uptake of informal or illegal mining activity by FARC ex-combatants, including the Nasa among them.

The first concern was easy to talk about. Most Nasa participants mentioned it outright. Pedro summed up the reasons why the Nasa were worried that FARC's retreat could pave the road for mining companies to enter Nasa territories:

> *For example, in Santander, there are several applications to extract gold and that is no secret. [...] Now, with the matter of peace and all what the post-conflict transition entails, I say for us there will be no peace, because armed groups will continue coming here. There is an interest in that. Perhaps the State itself, through the paramilitaries, will gain strength and they will be the ones who control the territory, the army itself. What is convenient for them is to get rid of one armed group and start introducing the entrepreneurs to extract all the different resources.*

Many others shared Pedro's concern that State interest in mining revenues will drive a wave of mining enterprise in Nasa territories, and that it will be accompanied and facilitated by right-wing illegal armed groups. The Indigenous people of Cauca have extensive experience in post-agreement environments, having witnessed the return to civilian life of several guerrilla groups that operated in their territories. In those same territories, Indigenous communities have been targeted by waves of paramilitary violence. They are aware that one illegal armed group can swiftly replace another as the controller of a territory. These fears are based on lived experiences of previous swings in territorial control between different violent groups.

The second concern, about intensifying illegal mining activity, related to the same question: Who would exercise control where FARC once did? FARC's demobilisation presented the participants with a more complex territorial scenario than previous demobilisations. When M19 and MAQL demobilised, FARC remained active for decades, exercising territorial control. That control, argued some, prevented new armed groups from emerging. In contrast, FARC reintegration would leave a vacuum of territorial control that could allow illegal mining to expand.

The third concern was harder for some to discuss because it involved hypothesising about what former FARC, including some Nasa, might do. For that reason, people were cautious in their statements. Some, like Samuel, simply said, referring to former combatants, "*We wonder whether they will seek to participate in mining*". Others, like Juan, were careful in their statements, but were likewise preoccupied. When I asked about any risks that the post-agreement might represent in relation to mining dynamics, Juan brought up former FARC combatants. He said:

> *When we go through a process of influencing people so that they hand themselves over, then they are under our responsibility. We follow a different process, so that the person gradually forgets the situation they had to live over there and gets used to our organisational dynamic. But when*

> *[the reintegration] isn't done this way, those other people who leave [the guerrillas] act that way, they look for easy money, they know how to use a revolver and all those things that start happening.*

So I asked:

> *Do you mean they are more prone to becoming miners?*

And Juan replied:

> *To some degree they are. Let's say it is not always the case, but yes, they can go into that, that is one of the ways. Likewise, they can go into illicit crops or other illegal options. It can also be mining, because they can think it's a viable option.*

While Juan cautiously excluded Nasa former combatants from the group of possible offenders, several other participants, like Samuel, did not. My impression is that this is a prevalent concern. The testimonies of miners in Chapter 3 suggest that ex-combatants do face difficulties making a living and might favour mining when they are otherwise not easily employable, lack the capital or land for agricultural work or aspire to higher incomes. Fears about contentious ex-combatant livelihood choices reflect Theidon's (2007) observations on the difficulties inherent to ex-combatant reintegration efforts in Colombia and on the need to look beyond the former combatant to consider community rights and expectations.

The many hopes of the ideal life of *Wët wët fizenxi*, with all the possible opportunities, became tempered with multiple concerns and with fears that the benefits might in fact be very difficult to realise; fears that what was more likely was a testing period of violence and illegal activity, including mining, in the aftermath of the Peace Agreement. Deaths in combat reached the zero point during 2016. There is great hope in that. However, according to the Colombian Ombudsman's statistics: i) 70 per cent of the 311 social leaders murdered in Colombia between 2016 and July 2018 did community work in rural areas; ii) they were murdered because they "*opposed two activities: drug trafficking and illegal mining*"; and 78 of these leaders were from Cauca (Cárdenas, 2018; Colombian Ombudsman, 2019). The statistics spoke for themselves. This brings us back to the ideas of positive (and negative) peace and to experiences that have led the Nasa to know that for as long as the underlying causes of social conflict remain unaddressed, the end of an armed confrontation cannot equate peace, it cannot guarantee negative peace, let alone positive peace.

Nasa Peace Efforts: Peacemaking and Peacebuilding

Nasa peace efforts touch all aspects of Nasa lives because, as a heavily conflict-affected community, the Nasa work to build harmony and protect rights, which are necessarily all-encompassing exercises. While acknowledging and celebrating the benefits of a ceasefire, the Nasa I interviewed always sought to go further. They saw their efforts to protect a wide scope of rights and gain redress as being part of the inventory of their work towards peace. In their repertoire of work towards peace, the Nasa ultimately aim towards positive peace, but also include many efforts towards negative peace, understood as the absence of violence, in the broad sense. As Galtung (2015) explained, negative peace is the absence of direct violence, of violence by omission (or exclusion) and of cultural violence that justifies the previous two.

The Nasa understand 'peacebuilding' as integrating activities that are treated separately in the academic literature – peacemaking, or negotiation efforts to bring an end to hostilities, such as in a peace process; and peacebuilding, or addressing the long-term issues and relationships between conflicting parties (Galtung, 1996; Ramsbotham, Woodhouse, & Miall, 2011). These categories prove artificial in describing Nasa peace efforts that are integrative and ongoing, regardless of national peace negotiations, and include regular negotiations of terms to mitigate hostilities in Nasa territories. For clarity, I shall refer to the Nasa peace efforts largely as peacebuilding, a broader category, unless I refer specifically to the negotiation of the Peace Agreement.

The narratives of some participants, such as La Mariposa, suggest that a peaceful resistance, or non-violent action, approach towards the armed conflict has contributed to shape today's Indigenous movement in North Cauca. In La Mariposa's narratives, the resistance to the armed conflict and to the systematic dispossession the Nasa have experienced has come to define the Indigenous movement in many ways. La Mariposa spoke of how:

> *Things are done against us and laws are passed that go against our rights, but that has made us organise ourselves as an Indigenous movement, to resist the war, not by taking arms, but with proposals, with debates, with protests, and by denouncing the facts.*

She maintained that the need to resist allowed the movement to create its structures and networks for life protection, self-government, justice, health and education, as well as the Cultural Territorial Plan. La Mariposa believed the Indigenous people have advanced not because of Government help, but because the Indigenous people have led "*an ongoing historic struggle that has cost them the lives of many people*". Acts of resistance have become part of an ongoing process where multiple acts build on each other. This, in turn, explains why the Nasa understand peacebuilding in a holistic fashion. The peaceful resistance has permeated all structures, plans and activities of the movement.

Many Nasa have lived in near crossfire conditions with the tangible presence of armed groups and combat, so failing to recognise day-to-day activities as peacebuilding would only reveal a poor understanding of the circumstances they have endured. In one of the first interviews I did during the fieldwork, I exhibited exactly that sort of superficial understanding. I was perplexed when a Nasa leader, Martha, assured me that the work I had seen her do during that day – making paper flags for a soccer tournament – was a peacebuilding activity. Trying to hide my surprise, and all too accustomed to the Colombian tendency to attribute soccer games almost magical powers, I enquired further: "*Why is the soccer tournament an act of peace, Martha?*" She smiled and said:

> *Compañera, the tournament is a peace activity because we are putting teams together and we are playing together, and for that we have to leave the past behind and just be able to play with our fellow community members. That is why this is an act of peace. We have to stop thinking that we are in a war. The war is over. We have to be a community again, leave the hatred behind.*

This made more sense to me, but I only came to gain a deeper appreciation of the meaning of Martha´s soccer flags when, in other interviews, I heard of the extent of FARC's Sixth Front presence in North Cauca Indigenous communities and of their recruitment from Nasa communities.

The Nasa have conducted extensive and wide-ranging peace efforts, including peacemaking and peacebuilding work. This work has looked inward (i.e. within the movement), and outwards to promote action nationally and internationally. I listened to detailed accounts and extensive lists of the activities of the Nasa Indigenous movement to contribute to peace. Andrés classified them in five categories. First, efforts to promote and raise awareness about the need for a negotiated solution, including advocacy, debate, internal awareness raising and education, protests, and denouncing and documenting human and collective rights violations. Second, establishing external alliances with those who could facilitate dialogue. Third, strengthening internal political and self-government capacity for peace. Fourth, developing a peace proposal from the Indigenous movement and articulating it into concrete Life Plans, programs and projects. Fifth, resisting the armed conflict through Indigenous Guardia actions and making this resistance visible.

Activities under the first category, actions to promote a negotiated solution, have been diverse. They have included direct demands for a peace negotiation, awareness raising efforts on the consequences of the conflict for communities, as well as deeper discursive engagement through what have been termed 'epistolary debates' with FARC's general secretariat. The frustration, fatigue and suffering through the continuous hindrance of everyday life by the war led the Nasa to push for a negotiated solution. La Mariposa described some of those frustrations and how they motivated the Nasa to call for a peace negotiation:

> *All of this led us to say there are two sides in this war. They don't let us work. They don't let our children study. They don't let us move. It is war all the time. We are tired. We said: 'Listen, sit down and talk. Stop the*

> *war. Why don't you talk? If you need a place, we can provide it' [...] Let's find a place but let's tell FARC and the Government to sit down and stop the war, because the war between them is affecting us.*

Nasa people have offered their peacemaking knowledge and skills. According to Juan, this has been done regularly and when the processes allow it. For example, where it was allowed in the Peace Agreement process. Juan said to me:

> *We have made practical contributions because, within the context of the conflict situation we have lived in our territories, we've had to go and talk to the armed groups and establish community agreements to, like they say, humanise the war. Obviously we have been victims of this war. So, we ought to minimise the harm.*

Nasa peace efforts have also encompassed awareness raising, advocacy and social mobilisation. According to Andrés, Samuel, Julián and several others, and as it is evident in ACIN's social media and the public record, the Indigenous movement documented and denounced the violence and rights violations that the parties to the armed conflict inflicted on the population. CRIC and ACIN have carried out large-scale and prolonged community protests to denounce and resist the conflict. In recent years, they have collaborated with the campesino and Afro-descendant communities in social mobilisations. This interethnic, intercultural work continues to the present day. The union across difference that Nicolás and Juan talked about is one of the strategies that the Nasa and other social organisations implement to raise awareness and exert pressure.

La Mariposa, Andrés, Julián, Cxayuce (pseudonyms) and many others recounted that the Indigenous movement promoted a negotiated solution through acts of collective, public, peaceful resistance guided by the motto "*count on us for peace, never for war*".[6] The movement raised national and international awareness of the Indigenous Guardia's role as defenders of life, of rights and of international humanitarian law, who resisted the actions of the armed groups, legal and illegal. In the 2000s, the Guardia achieved national recognition for its peaceful resistance and peacebuilding work (Redacción El Tiempo, 2004). The resistance to the armed conflict was also inscribed in the territory. Most Nasa settlements and infrastructure I visited have extensive collections of mural art with images of the resistance, often the Guardia or the Cauca Indigenous flag, and references to the life the community wishes to live, where agricultural motives, rituals, water and figures alluding to Mother Earth feature strongly.

The Nasa organisations have devoted research processes and rituals to documenting and remembering the victims of the conflict. Through collective, ritual territorial journeys, the Nasa commemorate the significant events of the armed conflict. There are countless examples. To name one, the *Casa de Pensamiento*, ACIN's think tank, as part of its memory of the conflict work, undertook a project to remember the victims. One of its outcomes was a "*walk through the territory*" that had ritual, performative and protest elements (González Rosas, 2016). It was held in March 2012 to remember Nasa victims of the armed conflict. Martha and Cxayuce told me that the march's name – "*Chiva, Memory and the Movement*" (La Chiva, la Memoria y el Movimiento) – referred to the Indigenous movement and the journey through the territory, made in a traditional Colombian bus or Chiva, to mark and commemorate Indigenous victims of the armed conflict in North Cauca. This event saw thousands of North Cauca Indigenous people walk through North Cauca marking the places where each of the victims fell. Martha told me that at each site there was a ritual and a speech, the planting of a rock in the ground, painted in CRIC's red and green flag colours and marked with the name of the victim. The families and communities of the victims reciprocated with performances and food for the participants. Some stops were for a single fallen leader, the perpetrators known or still at large. Other places, such as El Nilo, or San Pedro, were the sites of massacres that shook the Indigenous movement of the region (González Rosas, 2016). González Rosas (2016), who took part in the march, tells us in his ethnographic account of it that these families and communities profoundly valued these acts. They felt the sacrifice and tragedy had not been forgotten, after years or even decades. The act of walking through the territory is a convention in Nasa ritual, resistance and memory actions. González Rosas (2016) observed that these acts of resistance and assertion brought traditional spiritual practices up to date by reinterpreting them in the current political context.

The Indigenous movement's work needed allies to contribute to their peace efforts, so they approached community and human rights organisations and even sympathisers of the armed groups, who could facilitate dialogues like the epistolary debate with FARC's secretariat. Andrés told me that through that dialogue, the Indigenous movement made a strong critique of FARC that was very effective in raising awareness and gathering support. According to Andrés, that critique included, for example, telling FARC:

> *Well, you call yourselves defenders of the people, so why do you attack the people? You talk about democracy, so why do you attack the democracy that the community is building? You call yourselves revolutionaries but you are acting like new colonisers.*

This critique, Andrés believed, leveraged significant international support for the Indigenous people's campaigns for peace, and increased awareness of the implications of the armed conflict for Indigenous communities. The support of some international cooperation agencies, local and international NGOs, academics and students was palpable. I observed numerous groups of artists and students collaborating with the Nasa in designing and painting murals, delegations from universities taking part in rituals such as the Sakeluh, and delegates from multilateral agencies and NGOs making regular visits to follow up on projects, existing and proposed.

Andrés explained that there came a point where the Nasa organisations realised their need to strengthen community capacities to build peace. There was a gap between leadership discourse and grassroots capabilities to implement peace. Similarly, La Mariposa remembered:

> *[…] in 2015 we felt we had done sufficient outward advocacy work but we had a gap internally. The gap was that we let the community behind. So now we are doing that inwards advocacy. For example, what are those six points [in the Peace Agreement]? What progress has been made? What has been negotiated? There will be a plebiscite, so how will that affect us? We must vote on the plebiscite, so what will we propose in that context?*

As part of the internal strengthening, the Nasa devised a range of campaigns, activities and proposals to create the conditions for peace. They included debate, education and awareness-raising campaigns to deter Nasa people from supporting or joining any of the armed groups. One example that some participants, such as Julián, considered extreme was a decision to strip those who joined the ranks of FARC of their Indigenous status. The organisations designed programs and projects to promote economic activity, social development, culture and territorial protection. There were proposals to support Cabildos, as the highest territorial Authorities. For example, the Guardia, all of whom are connected to a Cabildo, were taking part in a conflict resolution school while I was in Cauca in 2016. Communities were receiving support to prepare for some of the most difficult and concrete requirements of peace, in particular of reconciliation. La Mariposa told me why this was so important. She said:

> *We are doing awareness raising and sensitising people within the communities to prepare ourselves for peace, for reconciliation. So, how is that going to happen? What about forgiveness? We are very worried because there are some who are still reluctant and we won't change that overnight […] For example with political participation: 'how is it possible that they have killed my people in the territory in my community and now they come to do politics and gain rights'. Should we give that to them or not? There are communities that say yes. But others say 'no, they must first pay and be accountable for what they have done, because they have caused a lot of suffering'. So we are working on those matters with people. Another one is re-socialisation, how will they link back into the territory? It is very difficult to get to all the 300 villages, but we are trying.*

The next level of the Indigenous effort was to plan for peace and formulate proposals to address the structural, systemic causes of the social conflict in Colombia. Andrés stated that the Life Plans that ACIN and its affiliated Cabildos created were essentially the development plans that the Indigenous community puts forward for that transition from the end of the armed conflict to a scenario of peace. Parallel to those Life Plans, Andrés expected the Government and FARC would have developed plans as well, and the question as to whose proposals are implemented would be one of legitimacy. Andrés was confident that the Indigenous movement was ready to contribute to peace "*be it with FARC as a political party, with the progressive left, with the government or even with the extreme right*". He felt that having worked on their own proposals for peace, and concentrating on addressing the structural causes of violence, prepared the Indigenous movement to contribute to "*building a new country*".

The Nasa inventory of peacebuilding work includes highly specific, territorialised resistance activities, where the Guardia adopts an instrumental role. This resistance is embedded in the territory, like when the Nasa declare a place 'a space of permanent assembly and a territory of peace'. La Mariposa recounted how, since 2002, with the arrival of the paramilitaries, the Nasa began to create such places:

> *We had to create permanent assembly places because, amid the conflict and the war, we had to concentrate and not let ourselves be pushed out of the territory and make sure they respected those areas. If they wanted to murder us, they'd have to murder us all, but not spread out.* [...]
>
> *There were eight large permanent assembly places [...] but besides there were other smaller ones like the schools and health centres where we placed the flag, and when we gathered together. If there was a crowd, we would turn wherever we managed to get to into a permanent assembly place.*

The strategy with those assembly places was to gather in large numbers and protect those places from the armed actors and from combat. Some were pre-established, some emerged out of immediate need as the armed conflict events occurred.

In 2011, this direct resistance became even more decisive. La Mariposa recalled:

> *We said, we'll drive them out of our territory, army and guerrillas, we will drive them out [...] we did such things, that we wondered, should we leave it as a threat or do it? We told them, you are not welcome. Later we said: if we have to go to their camps we'll go, but they, army and guerrilla, have to leave. We took the Berlin mountain, later the army would take control of a community and we would go and drive them out, the guerrilla would take hold of a community and we'd push them away. We let them go through but did not let them stay. So the situation was such that we became enemies of the army and of the guerrilla. So the guerrilla started to kill leaders because we did not leave them alone.*

The Defence of Life, and Justice and Harmony networks led this confrontational resistance as well as the reparation and healing work with victims.

Defence of Life was in charge of protecting human rights through the Guardia. Justice and Harmony performed investigations and sanctions based on Indigenous justice and dealt with the Indigenous justice–mainstream justice interface. These networks worked to protect victims' rights and provided investigative backing for cases to demand redress.

Nasa Peacebuilding and Responses to Mining: Articulating Notions of Peace, Harmony and Resistance with Moral Views on Mining

Most Nasa participants told me that articulating a response to mining is a form of peacebuilding. Their views on the mining-peace relationship brought together their ideas of peace and their moral views of, and experiences with, mining. Overall, the participants thought that mining scenarios that did not resonate with their moral views, or that threatened their livelihoods, were obstacles to peace. The notion of peace is strongly grounded on morality and material circumstances. This relationship between peace, morality and the material conditions of life surfaces in a range of opinions about the mining-peace interaction. Several participants saw any form of mining as an obstacle to peace. Others believed only mining by outsiders or that uses chemicals, heavy machinery or violence hinders peace. Another group stated that obstructing the work of Indigenous miners, some of whom are ex-combatants, fuels conflict by depleting their livelihood. There were intra- and inter-community conflicts about mining and where participants located themselves in the spectrum of those conflicts informed what responses to mining they saw as peacebuilding. This way, some participants saw stopping all forms of mining as peacebuilding, others thought blocking the entry of foreign miners was peacebuilding, while some believed that it was regulating mining, not banning it, that would build peace.

Rather than implementing programs to address intersecting elements of mining and post-conflict peacebuilding, the Economic-Environmental, Defence of Life, and Justice and Harmony networks addressed different aspects of the problem. However, that did not mean that there had been an analysis or a mandate to respond to identified risks or opportunities in the intersection between mining and post-conflict dynamics. Specific questions such as the engagement of former combatants in mining in the post-agreement period were concerns amongst the Nasa, but there were no targeted responses in this area at the time I was in North Cauca. A predominant view amongst the participants was that responding to mining is building peace. Within this there was consensus – all but one participant agreed – that stopping external miners, from 'the multinationals' to the national companies, from entering Nasa territory contributes to harmony and to the freedoms of the Nasa people that are intricately linked to Nasa ideas of peace. However, there was less clarity as to what peace constitutes when it comes to Nasa miners. This is what I shall concentrate on in the following paragraphs.

What peace, harmony or peacebuilding are can vary according to perspective when contested forms of mining are concerned. Many participants, like

Johana (pseudonym), a practitioner from ACIN, could not understand why miners persisted. She believed that to protect the territory mining should be stopped. On the other hand, Botón de Oro (pseudonym), a miner, feared that the ban on mining could reinvigorate community conflict or leave former FARC combatants without viable livelihood alternatives. The Nasa miners recommended dialogue between Cabildo and mining leaders to bring peace, while participants with experience in research or dialogue about mining, or with mining itself, thought that rather than a blanket NO, a nuanced response grounded in research and dialogue was necessary to address the causes of the conflict and the divisions surrounding mining.

Given the diverging ideas of the peace-mining relationship, Nasa responses to mining can be understood as peacebuilding in light of some perspectives but not of others, so whether they actually effect peaceful relationships is unclear. Let us concentrate on a broad definition of peacebuilding such as in Ramsbotham, Woodhouse and Miall (2011), where peacebuilding is what addresses structural, long-term issues and relationships between conflicting parties. From the varying perspectives of the participants, the responses to mining they suggest or support can all constitute peacebuilding. For example, based on the moral understandings outlined in Chapter 2 and earlier in this chapter, acting to halt mining is a way of starving armed conflict of resources and of preventing the processes of social, cultural, moral, environmental and economic deterioration that hinder harmony and peace in communities. In parallel, blocking a mining tunnel without addressing the livelihood questions of the miners would fail to address the structural issues that have led to mining and to the conflict around it. For some miners, the prohibition goes against their notion of peace, which is strongly centred on livelihoods. As a result, some of these responses might be construed as peacebuilding by one party, but when seen from another party's perspective the same responses might be construed as feeding conflict. Only the responses to mining that are informed by the perspectives of all parties to the mining conflict and that make targeted attempts to address structural dynamics are likely to contribute to peacebuilding.

Livelihood solutions emerge as the most prominent example of a response to mining that can contribute to building peace, but their effectiveness depends on other identity and morality centred work. The programs to promote agricultural livelihoods were the subject of discussion in Chapter 4. In that discussion, testimonies from ACIN and CRIC leaders made it clear that those programs faced serious budgetary limitations. Where there is difficulty in sourcing the funding for alternative livelihood projects, it is uncertain whether such projects could compete with the substantial windfalls that some attribute to mining. To tilt the balance away from the kind of uncontrolled, polluting mining with links to violent groups that the Nasa fear, other work was necessary. Here efforts that seem less targeted at economic gain gather importance. Activities to promote pride in an agricultural identity, promote the value of moral standing in the community as a source of self-worth or reassert stewardship responsibilities over water and the reciprocity relationship with Mother Earth are some examples.

What happens where agriculture is not a viable or sustainable option for people or families? Where agriculture is a viable alternative, the view that preventing mining is peacebuilding is a coherent one. In that scenario, actions to protect and expand, or recover, Indigenous territory and to ensure environmental care, by expelling miners including Indigenous ones, contribute to peacebuilding. But the discussion becomes more complex when we consider the limitations that agricultural livelihoods are facing in some Nasa Resguardos. It is clear from the testimonies from ACIN leaders and some miners that land availability and population growth are posing barriers to sustainable agricultural livelihoods. This is driving, for example, the 'Liberation of Mother Earth' campaign. This is how an environmental specialist explained the situation to me:

> *In Canoas – even though they are very hardworking – there's a lot of us. We are 9,000, or 2,800 families already. As far as land tenure goes, we are very tight. There are families with one hectare and five children. So if you divide 10,000 square metres between five people, what would that look like? So our policy since the 2000s has been to buy little farms and give people one or two-hectare parcels, but it is not enough. So the Indigenous congresses have mandated that the first priority is to recover land, now we call it the Liberation of Mother Earth campaign. We have done it in several Resguardos to recover land and to give it to families who don't have it. […]*
>
> *That has an environmental aspect: In Canoas and some other territories there are environmental codes or rules. All of those say that the water holes must be targeted with reforestation programs and should not be disturbed so that water flows are maintained. But because there is no land, there are families that go right up to the water hole and start chopping down trees. That has been a battle here in the north. So we decided that for as long as some have so much land to plant sugar cane and also benefit from the environmental care we do in the mountaintops with the water sources, then we should remain in the Liberation of Mother Earth process. The State proposes to continue buying land to relocate Indigenous people to the mountaintops. But then that means we cannot preserve the higher reaches of the mountains. But in our Resguardo, we have bought land to plant trees. We bought 90 hectares in the Cerro [Munchique]. It had 17 large water holes. We bought it and left it alone. Today those are tall forests that we have in the Resguardo to preserve water.*

The land scarcity situation of some Resguardos comes from the interplay of broader economic choices and policies, and was placing limits on agricultural livelihoods, while driving environmental degradation. In Canoas, the latter had been addressed, however, the lack of arable land for community members remained. As this participant explained it, the land situation spoke of societal choices about land distribution and about the preferred model of rural development. The Nasa oppose sugar cane and other monoculture agroindustries and favour the family-based farm model. However, successive governments have legislated in favour of industrialised agriculture. Ultimately, the ongoing relocation of Indigenous people to the mountaintops is a contemporary iteration of an historic process of economic displacement that has been discussed by others such as Rappaport (2000) and González Rosas (2016).

The question of peace, in particular as it relates to mining, is intricately linked to the question of livelihoods and needs further study within the Nasa communities. Those with access to land, or with the networks and community standing to obtain it, are best placed to be agriculturalists. Once we consider this, the discussion about whether stopping mining leads to peace starts moving beyond the presence of illegal armed groups and the risk of environmental degradation, and into the realm of power dynamics and existing networks of privilege within Nasa communities. The moral tension over mining reflects an aspect where the community was not at peace, where there were political and economic contests. Some people were unable to make a living from agriculture. Some might have aspired to more material wealth and possibilities than agriculture allowed. How can the Nasa establish a clearer path for their responses to mining that can acknowledge the difficulties of all sides and bring community harmony? This will be difficult to achieve for as long as there are no reliable surveys of Nasa miners, information on any illegal armed groups involved, nor detailed analyses of land ownership's correlation with mining uptake. Most of these remained to be completed at the time of the study. Political will, physical danger and funding restrictions had precluded the necessary investigations.

In the meantime, the risk of violence and clashes between supporters and detractors of unsanctioned mining will remain, but the risk is not exclusive to mining, nor to the Nasa. Instead it reflects structural problems of economic inclusion in Colombia. As miners continue their enterprise in Indigenous territories, they are likely opening or maintaining channels of control for illegal armed groups. For as long as the land access, livelihood and power distribution dynamics remain unchanged, the miners are likely to persevere. If the threat of illegal armed group control persists, the Nasa authorities' ban will likely remain. The clash between the two parties will likely endure and one can only wonder: When will the gold price drop enough that mining it is not as attractive? But in that scenario, as Luis (pseudonym) explained earlier, there is a perverse dynamic where families facing financial difficulty swing between the illegal economies of mining and drug production and trafficking. Furthermore, all these questions connect to the broader national question of rural development, one that invites very different responses from the Nasa and from the National Government. The Nasa did not have all the resources to address what some would call an illegal mining crisis. They persisted in resisting it, but it was clear that it placed lives and community harmony in danger. So what is the role of the State and how can it best liaise with the Nasa? The Nasa maintain they do not have sufficient land and they oppose the expansion of monoculture agroindustry. In a region with high concentrations of rural Indigenous populations who support alternative models of rural development, how can policy be designed in a more inclusive manner? Ultimately, the answers about mining in the post-conflict period relate to the broadest questions that have motivated political contentions, including violent confrontations in Colombia historically. Given that mining was not directly addressed in the Peace Agreement, the responses would likely be devised along the way. At the time of the study the scenario was not encouraging.

Conclusion

In this chapter, I have concentrated on the relationship between peacebuilding and responses to unsanctioned mining in Indigenous territories, which was the form of mining causing most divisions within the Indigenous movement at the time of the study. Where mining by outsiders was concerned, there was broad consensus that resisting mining equated to building peace. There was a prevalent view among the Nasa that there were connections between mining and illicit economy networks with links to violent armed groups. The Nasa resisted mining, which they saw as an activity that built on and strengthened existing infrastructures of violence. This resistance followed similar motivations and processes to those of the peacebuilding activities that the community is known for. Through peacebuilding and resistance to mining, Nasa organisations sought to protect their political and economic interests and the rights of their communities. However, livelihood dilemmas, aspirations for better material conditions and political contests saw some Nasa participate in unsanctioned mining ventures in Indigenous territories. In these scenarios, the Nasa were divided on what constitutes peacebuilding where responses to mining are concerned. Ideas about harmony and what is morally appropriate, as well as individuals' positioning within the mining conflict, in particular as it relates to livelihoods, informed what they considered peacebuilding in relation to mining. This reflected an internal conflict within the Nasa community. It mirrored the contours of other internal disputes about the illicit economies that fuelled the armed conflict such as the drug trade.

Despite the divisions, Nasa organisations have responded to the dynamics of the armed conflict and of mining with a repertoire of programs and actions. Those actions have faced the problems arising from the illicit economies of drugs and mining in a similar fashion. However, significant aspects remained unaddressed and were not within the full control of the Nasa. Illicit economic activity continued to pull in members of rural communities like the Nasa who, despite the well-known risks and the documented connections with violence and the armed conflict, saw in mining or 'crops of illicit use' a way out of poverty. Ultimately, violence creeps in through the cracks left by dispossession and material poverty and, as long as Colombian society fails to attend to those problems in rural territories such as North Cauca, communities' best efforts will continue to fall short.

Notes

1 With different legitimate types of authority that "co-exist, compete, overlap and blend" (Boege, 2010, p. 1) and with hybrid forms of legitimacy.
2 In the sense the peacebuilding literature uses the term liberal peace.
3 Rappaport (2000) uses Bourdieu's (1977) definition of *habitus* as a set of complex dispositions that make the members of a society favour certain practices over others. Rappaport (p. 222) states that habitus relates to experience and to the past and, in the Nasa case, it "includes, amongst other elements, the oral and written memory of *caciques*, the practical memory of ancestors experienced through ceremonies and the geography and commemoration of resistance towards the State at the personal level" (my translation).
4 Or accumulation by dispossession.
5 The Guardia.
6 See ONIC (2014).

References

Boege, V. (2010). How to maintain peace and security in a post-conflict hybrid political order: The case of Bougainville. *Journal of International Peacekeeping*, 14(3–4), 330–352. doi:10.1163/187541110X504382.

Cárdenas, H. S. (2018, July 5). *El mapa de los 311 líderes asesinados en Colombia El Colombiano*. Retrieved from www.elcolombiano.com/colombia/mapa-de-lideres-y-defensores-asesinados-en-colombia-DI8956261.

Chandler, D. (2013). Peacebuilding and the politics of non-linearity: Rethinking 'hidden' agency and 'resistance'. *Peacebuilding*, 1(1), 17–32. doi:10.1080/21647259.2013.756256.

Chapagain, B., & Sanio, T. (2012). Forest user groups and peacebuilding in Nepal. In P. Lujala & S. Rustad (Eds.), *High-value natural resources and peace building* (pp. 561–578). London: Earthscan.

Colombian Ombudsman. (2019). El riesgo de los defensores de derechos humanos merece mayor atención del Estado. *Defensor*. Retrieved January, 2019, from www.defensoria.gov.co/es/nube/noticias/7716/%E2%80%9CEl-riesgo-de-los-defensores-de-derechos-humanos-merece-mayor-atenci%C3%B3n-del-Estado%E2%80%9D-Defensor-Defensor-del-Pueblo-Carlos-Negret-Defensor%C3%ADa-derechos-humanos.htm.

Consejo Regional Indigena del Cauca (CRIC). (n.d.). *Puntos de Cambio del Programa de Lucha*. Retrieved July, 2017, from www.cric-colombia.org/portal/estructura-organizativa/plataforma-de-lucha/.

Cramer, C., & Wood, E. J. (2017). Introduction: Land rights, restitution, politics, and war in Colombia. *Journal of Agrarian Change*, 17(4), 733–738. doi:10.1111/joac.12239.

Escobar, A. (1995). *Encountering development: The making and unmaking of the third world*. Princeton, NJ: Princeton University Press.

Galtung, J. (1996). *Peace by peaceful means: Peace and conflict, development and civilization*. London: Sage Publications.

Galtung, J. (2015). *International encyclopedia of the social & behavioral sciences*. Waltham, MA. Elsevier Ltd.

González González, F. E. (2014). *Poder y violencia en Colombia*. Bogotá, Colombia: Odecofi-Cinep.

González Rosas, A. M. (2016). *Vivimos Porque Peleamos – Una mirada desde abajo a la resistencia indígena del Cauca, Colombia*. Ciudad de México: Memorias Subalaternas.

Grajales, J. (2011). The rifle and the title: Paramilitary violence, land grab and land control in Colombia. *Journal of Peasant Studies*, 38(4), 771–792. doi:10.1080/03066150.2011.607701.

Jabri, V. (2013). Peacebuilding, the local and the international: A colonial or a postcolonial rationality? *Peacebuilding*, 1(1), 3–16. doi:10.1080/21647259.2013.756253.

Lederach, J. P. (1995). *Preparing for peace: Conflict transformation across cultures*. Syracuse, NY. Syracuse University Press.

Lujala, P., & Rustad, S. (2012). Part 5: Livelihoods: Introduction. In P. Lujala & S. Rustad (Eds.), *High-value natural resources and peace building* (pp. 463–466). London: Earthscan.

Mac Ginty, R. (2008). Indigenous peace-making versus the liberal peace. *Cooperation and Conflict*, 43(2), 139–163.

Mac Ginty, R. (2010). Hybrid peace: The interaction between top-down and bottom-up peace. *Security Dialogue*, 41(4), 391–412. doi:10.1177/0967010610374312.

Mac Ginty, R. (2015). Where is the local? Critical localism and peacebuilding. *Third World Quarterly*, 36(5), 840–856. doi:10.1080/01436597.2015.1045482.

Mac Ginty, R., & Richmond, O. P. (2013). The local turn in peace building: A critical agenda for peace. *Third World Quarterly*, 34(5), 763–783. doi:10.1080/01436597.2013.800750.

ONIC (National Indigenous Organisation of Colombia). (2014). *Agenda Nacional de Paz de los Pueblos Indígenas de Colombia*. In O. N. I. d. Colombia (Ed.), (p. 2). ONIC.

Pugh, J. (2018). Weaving transnational activist networks: Balancing International and bottom-up capacity-building strategies for nonviolent action in Latin America. *Middle Atlantic Review of Latin American Studies*, 2(1), 130.

Ramsbotham, O., Woodhouse, T., & Miall, H. (Eds.). (2011). *Contemporary conflict resolution*. Cambridge, UK: Polity Press.

Rappaport, J. (2000). *La Politica de la Memoria*. Popayan, Colombia: Editorial Universidad del Cauca.

Redacción El Tiempo. (2004, December 7). Premio Nacional de Paz para Indigenas y Cacaoteros, *El Tiempo*. Retrieved from www.eltiempo.com/archivo/documento/MAM-1525605.

Rettberg, A., & Ortiz-Riomalo, J. F. (2016). Golden opportunity, or a new twist on the resource-conflict relationship: Links between the drug trade and illegal gold mining in Colombia. *World Development*, 84, 82–96. doi:10.1016/j.worlddev.2016.03.020.

Richmond, O. P. (2010). Resistance and the post-liberal peace. *Millennium – Journal of International Studies*, 38(3), 665–692. doi:10.1177/0305829810365017.

Richmond, O. P. (2015). The dilemmas of a hybrid peace: Negative or positive? *Cooperation and Conflict*, 50(1), 50–68.

Scott, J. C. (1985). *Weapons of the weak: Everyday forms of peasant resistance*. New Haven and London: Yale University.

Taussig, M. T. (1984). Culture of terror—space of death. Roger Casement's Putumayo Report and the explanation of torture. *Comparative Studies in Society and History*, 26(3), 467–497.

Taussig, M. T. (2004). *My cocaine museum*. Chicago: University of Chicago Press.
Theidon, K. (2007). Transitional subjects: The disarmament, demobilization and reintegration of former combatants in Colombia 1. *The International Journal of Transitional Justice*, 1(1), 66–90. doi:10.1093/ijtj/ijm011.
Tsing, A. L. (2005). *Friction an ethnography of global connection*: Princeton: Princeton University Press.

8 Perspectives from Outside the Nasa Organisations

Introduction

I have placed the perspective of the Nasa at the centre of this analysis and now it is time to look outward to contextualise the Nasa discourse within the tapestry of diverse views about mining, conflict and peace in Colombia. The 'views from outside' that I shall discuss here, rather than being a representative sample, are examples of the kinds of perspectives one can find in Colombia on these matters. The participants came from sectors of society where views are not always unified either. In analysing their perspectives, my purpose is not to measure or compare them in scales of truth, but to delineate the contours of a landscape of diverse understandings. This is necessary because it is in that landscape that the conflicts and the problems about mining, and possibly the solutions and the collaborations, take shape. The contours will come mainly in the form of confluences and divergences about the role of the State in mining governance and in peace, and about the interactions of the State with Colombian regions and their communities. I shall discuss how the views of diverse actors, expressed in discourses about livelihoods, human rights, collective rights, sustainable development and social responsibility, converge in many respects with the concerns of the Nasa. Later, I shall turn to the central divergence – a clash between those views and discourses of efficiency and mining-driven economic growth.

I interviewed 22 people, including members of Afro-descendant and campesino organisations, specialists and researchers who have analysed mining in Colombia extensively, people from multilateral and international aid agencies, State officials, and people working in the formal mining sector in Colombia. Forty-one per cent, or nine, of the participants were women. There were three campesinos, one Afro-descendant, four international cooperation or international observers, six State representatives, five specialists or academics, and three industry participants. Although they did not always speak from an institutional perspective, it is important to locate institutional affiliations and, where relevant, I refer to those in the text. I interviewed participants affiliated to Cauca Afro-descendant and Campesino organisations, namely ACONC, FENSOAGRO and ANUC; the United Nations (UN), the United Nations

DOI: 10.4324/9781003226895-8

High Commission for Human Rights in Colombia, and the USAID Human Rights Program in Colombia; the National Mining Agency, the Mining and Energy Planning Unit, the Cauca Natural Resource Management Authority, the Office of the High Commissioner for Peace, the Cauca State Attorney; a multinational mining company, the Dialogue Group on Mining in Colombia, and the Colombian Mining Association; Colombia Punto Medio, the Regional Centre on Responsible Business and Enterprise (the Spanish acronym is CREER), Los Andes University and the Peace Research Institute Oslo; as well as an independent Colombian ethnic rights expert. Because of the small number of people, I will refrain from stating the institutional affiliation of the community participants to avoid identification.

Confluences

Considering how polarised Colombian society is, I found a surprising degree of confluence in the concerns and views of diverse participants from outside the Indigenous movement. In analysing their views, it became clear that their concerns gravitated around questions related to the role of the State in mining governance and in peace. The concerns of these participants related to four problematic areas: first, problems related to the way the Government understood mining that permeated into governance and policy; second, problems within the legal framework about mining and rights, and with its interpretation and application; third, problems resulting from a poorly conceptualised transition away from illegal economies in the post-agreement period; fourth, problems emerging from the pervasive gaps in State presence. For the participants these problems raised serious concerns about human rights in the post-agreement transition, or about the context for existing and future mining enterprises.

There were strong confluences between the views of representatives of other rural communities, such as the Afro-descendants and campesinos, and those of the Nasa about mining. There was resonance in the way they understood the links between mining and armed conflict dynamics and in their fears about the post-agreement transition. Rural leaders expressed apprehension that social conflict could intensify in the post-agreement transition if external development models were imposed on rural communities. There was coincidence in opposition to large-scale mining and mining by people external to the communities. This allowed for collaborative responses between Indigenous, Afro-descendant and campesino communities, under the understanding that ancestral gold mining is part of the Afro-descendant economic matrix and that communities mine essential materials such as ballast. Mining did not appear to be a primary activity in campesino communities in the region at the time of the fieldwork, but rather a source of tension with outsiders that attracted resistance.

The members of rural community organisations drew similar links between mining and armed conflict as the Nasa did. Most rural community participants had similar understandings and were particularly concerned about the involvement of criminal bands because of their brazen proclivities for violence. A participant stated that:

> *There is a direct relationship with FARC. I don't know what happens now with the process ahead with FARC, but in some places it has its own mine, in others just extorts. Same with the ELN and the paramilitaries, who also have their own illegal mines and in some cases extort payments and are responsible for transporting the produce from the mine to Cali or Medellin. There are many places where the illegal mining business belongs to armed conflict actors. […] The other thing is that the majority of threats when we do territorial control with the Indigenous Guard and the Cimarron [Afro-descendant] Guard come from paramilitaries – the Rastrojos or Aguilas Negras bands.*

The convergence extended to understandings of post-agreement scenarios in relation to mining, with participants expressing apprehension that the Peace Agreement implementation might lead to more mining agent entry attempts in post-conflict territories and that this could lead to increasing violence and repression from illegal or legal actors.[1] It was important for these participants that communities be consulted and that diverse views of development be able to coexist. The fact that there were similar views on these matters facilitated general collaboration around mining amongst these communities.

Problems in Understanding Mining

Specialists, academics and members of multilateral and aid organisations were concerned that the Government understanding of mining was inaccurate because it did not embed human rights and sustainable development knowledge, nor did it integrate practical lessons from the Colombian experience. To these participants, the approach of the Government to mining was biased, because it concentrated merely in formalisation and foreign direct investment growth. In doing so, it was overlooking the well-documented (CINEP, 2012; de Angulo et al., 2016; Moor & Sandt, 2014; Negrete, 2013) risks to human rights from mining, legal or illegal. Several participants thought the Government was yet to recognise that formalisation does not directly address the problem of illegality and that there are complex interactions between legal mining and both the armed conflict and illegality. These participants were joined by one from Government in favouring an understanding of mining as a social dynamic and as a rights matter rather than simply a formalisation matter, and called for a territorial approach to the problems with mining to give voice to local people and knowledge. Interviews with three Government participants indicated that while there were pockets of awareness, there were difficulties gaining traction with those approaches in Government mining institutions.

A participant from the multilateral sector, working in the human rights field, stressed the dire human rights situation of the communities in North

Cauca, which highlights how central human rights knowledge is in understanding mining in the post-agreement period. This participant explained to me that FARC militants in North Cauca stayed in their territories rather than joining FARC in other regions. Therefore, victims and perpetrators can live in the same territories. In Cauca, the participant told me, FARC's drive to exert higher levels of control had met strong resistance from Nasa leaders and Traditional Doctors, resulting in the murder of many of them. The situation would not become simpler with the retreat of FARC, according to this participant. Just like the Nasa, the human rights expert was concerned that, as a large source of employment and income based on the illegal economies of coca, marijuana and mining, FARC would leave a power gap that others were rushing in to fill. Violence was growing with FARC's retreat, intensifying human rights risks.

Several experts emphasised that human rights risks apply to both illegal and legal mining and mining formalisation does not mitigate human rights risks. "*The right to a healthy environment is a human right,*" said a UN official, who was concerned about the health risks from legal and illegal mining for nearby communities and downstream water users, including large population centres like Cali. The official stated that illegal and legal mining have put at risk the social, economic and cultural rights of the Indigenous people but also of the Afro-descendant people, whose artisanal mining traditions were being displaced by new mining operations. Several participants stated that formalisation does not address human rights risks because: i) prior to formalisation, mining operations might have infringed on human rights; ii) the formalisation process does not include human rights criteria; iii) frequently mining title applications, or titles, and illegal mining operations overlap; and iv) the status of legality is not a guarantee for human rights.

Based on the multiple interactions between extractives, illegality and armed conflict, a participant from the academic sector shared the scepticism about the idea that formalisation is a silver bullet. Interactions between large-scale mining or large-scale extractive activity and the armed conflict have been varied, according to this environment and sustainable development expert. This participant described several ways large-scale mining or extractives interact with the armed conflict. One way was through extortive practices. There is a strong precedent of extractive activity extortion payments contributing to strengthening armed groups (Molano, 2015; Verdad Abierta, 2015). The participant cited the extortion payment ELN obtained in the 1980s from an oil pipeline construction company that contributed to the strengthening of a languishing guerrilla front in Arauca. Another cited interaction was direct participation in the armed conflict (see also Moor & Sandt, 2014). A further modality was coexistence, where companies live side by side with significant illegal mining activity in regions with a history of violence, such as Mineros S.A. in El Bagre. This array of examples highlighted that formalisation was no panacea when it came to addressing links between mining, illegality and armed conflict.

As regards to illegal mining, a human rights expert was concerned about the gender violence and child molestation problems in illegal mining enclaves, and the chain of services built around those enclaves, which broke the social fabric of the communities. "*The income for the owners of the machines is significant, it is not so for the workers, but it is higher than what they could earn otherwise, so if someone wants to stop these illegal mining operations, the participating community members will defend them,*" said the official. In specific cases in North Cauca, female leaders from those areas have had to leave their territories because "*it was clear that they were going to kill them,*" said the official. Illegal mining poses significant human and collective rights risks.

A UN official stated that it was necessary for the Colombian Government to begin understanding mining as a social dynamic and a matter of rights rather than as an issue to negotiate through legalistic processes or a mechanism for territorial control. The concern was that new economic agents were attempting to force mining into territories, where it could be used as a territorial control mechanism. "*It is normal that ACIN would feel a lot of fear that, when FARC is no longer there, lots of economic interests will seek to enter, so it is important to define protocols for this,*" said the official. The risk extended to the process of formalisation. To this participant, the formalisation process should include in its criteria the viability of the operation and its ability to guarantee rights.

For several participants, including one from Government, it was clear that human rights and territorial approaches were necessary elements in a response to the problems with mining. For example, a UN official stressed the value of locally driven solutions because they are more likely to support rights and respond accurately to local dynamics. Part of the solution to the problems with mining, according to another UN official, was to strengthen community driven processes and governance. This representative highlighted the importance of supporting the work on Indigenous rights that the Nasa community do, stating that "*the Nasa are a very hard working Indigenous people, their work on rights, their proposals, their negotiations end up benefiting all the Indigenous people elsewhere in the country*". Meanwhile, an official from the Mining and Energy Planning Unit saw the need for a change in mining policy-making through a rights approach and a territorial approach that could respond to the different needs of communities and territories. This would be a strong departure from the one-size-fits-all that several studies have attributed to mining policy (de Angulo et al., 2016; Viana, 2015) and would align with the approach the UN officials suggested.

A Problematic and Poorly Implemented Legal Framework

Several participants thought that the Colombian legal framework on rights and mining had problematic aspects in theory and in practice.[2] The concerns related to poor implementation of FPIC; a precarious context for community participation; the division between soil and subsoil rights; and regulators' inability to balance rights and local considerations with foreign direct investment promotion. In this context, there was scepticism about Government capacity to respond effectively to the expected shift to higher rights recognition for campesinos as a collective in Colombia.

A UN official was concerned that mining titles are regularly granted without community consent, thus violating collective rights. The participant found the FPIC process in Colombia wanting and stated:

> *In ethnic territories, because of the law and because of International Standards, a process of free, prior and informed consultation and consent should be implemented and people here forget the consent. It is not just asking, getting NO as a response and then just getting on with the project. What it means is that if they say NO to you, then you must at the very least go back to the drawing board and change the proposal. It is also about the way you engage with the community and how you make the consultation process. There are huge weaknesses here [...] This is a human rights problem and the development model, in order for it to be sustainable development, cannot be done without respecting free, prior and informed consultation and consent.*

A UN representative was concerned that the allocation of soil and subsoil rights hindered consent processes and negotiation conditions towards a sustainable development. This official believed the legal excision of subsoil from Indigenous land title was incoherent and damaging, observing that there are jurisdictions where First Nations land titles include the subsoil and the government has the responsibility to address the negotiation imbalances between mining companies and Indigenous communities.

These concerns about poor standards of participation and rights safeguards resonated with two interviewees from the mining industry. One of them stated that the Government did not seek input from the ethnic peoples of Colombia for its decisions on mining, despite their territories covering 30 per cent of Colombia and correlating highly with the location of valuable minerals. Another went further and conveyed apprehension about the "*stigmatising and repressive tendencies of this Government [President Duque's] against community protest*". These participants saw the poor participation standards as contributing to the conflicts and polarisation of the mining debate in Colombia.

The investment sector itself was calling for the integration of human rights considerations in the regulatory framework for business. A participant from the multilateral sector had witnessed international businesses and investors manifesting a need for Colombian legislation to set the parameters for compliance with the international standards on business and human rights, and for a State capable of balancing the participation in international markets with the protection of rights. The participant stated that foreign investment is "*a matter of rights, of access to goods, of access to work, of labour rights that involves the private sector and the State*" rather than only a matter of regulating for the benefit of overseas companies or investors.

But the difficulties ahead for a transition into sustainable peace did not end with the obligations of the State. A human rights expert working in the international aid sector stated that, in the post-agreement, "*One of the biggest challenges for all of us and for Indigenous communities as well is going to be to recognise the emergence of campesino communities' collective rights, including their right to land*". The expert stressed that internationally the campesinos are acknowledged as having collective rights and that the time would soon come in Colombia where their culture and economic practices would be protected, with a requirement to consult these communities collectively being one of the consequences. A participant from the campesino movement explained that the pursuit of this recognition would continue to guide their efforts:

> *We will continue with the same actions, because things have not changed, for example, the work to gain legal, political, cultural and economic recognition. If the small-scale production culture is not recognised then they will continue to say that they need monoculture, with specific inputs and certified seeds. That is against the effort to maintain certain seeds and certain species that have been important for survival [...] that people don't have to buy.*

Problems in the Approach to Mining-Illegality-Armed Conflict Interactions

There was a generalised concern that the Government made no effort to understand the links between mining, illegality and the armed conflict and that as a result the response to those links had been poorly designed. Several participants suggested that the Government was not interested in effecting any change to the way mining was run in Colombia as part of the post-agreement transition. The Government's priority was to pass reform to 'unblock' the advance of mining. A specialist with in-depth insight on the peace process observed a limited degree of interest from Government mining institutions in understanding the implications of the Agreement for mining, other than as a mechanism to fast-track reforms that would otherwise require a prior consultation process. As a result, where the Peace Agreement devoted a chapter to the problem of drugs, it did not address the question of mining formalisation or alternatives to mining livelihoods. Several participants noted there were weaknesses in the design of the transition away from illegal economies, gaps in the mechanisms for land restitution and inappropriate programs to respond to illegality. All of these have implications for responses to mining-illegality-armed conflict interactions.

One participant explained to me that there were logical entry points to address mining themes as part of the Peace Agreement negotiation and implementation, but directives to deal with illegal mining or other mining-related concerns were not established, leaving the burden with the affected communities. This interviewee explained that there were thematic avenues to discuss mining in the Peace Agreement in its point one, about holistic rural reform, and in point three, about the cease to hostilities. Point one dealt with the formalisation of the rural sector, which is relevant to mining formalisation. Point three included the end of extortive and illegal practices that is likewise relevant to mining. But overall, to this participant, the National Government's position to cordon off the economic model from the negotiation left mining largely unaddressed.

The same person had heard the complaints and concerns of community members about the post-agreement transformations and their interactions with territorial mining dynamics. A significant worry for many was the military control vacuum left in former FARC areas of influence that allowed new agents to engage in military contests, and existing and new agents to seek political influence in communities in the mining frontier. In addition, the legal mechanisms for land restitution as redress for conflict-related land dispossession did not consider avenues to return land to those who lost it to legal mining interests. As a result, most complaints related to illegal mining, where routes for land restitution were available to communities with collective title. Furthermore, the participant spoke of likely conflicts that would emerge in areas prioritised for rural reform as part of the Agreement if they had mining wealth but no mining vocation.

For one participant from the UN, a significant source of risk for the rights of communities and for territorial peace was the poor design of the transitions away from illegal economies within the Peace Agreement, heightened by additional complexities for the case of illegal mining. Part of the problem was that FARC was meant to retreat before economic transition was set in motion in its areas of influence. This was problematic to the official because FARC had the relationships with communities connected to those illegal economies and if FARC exited before a transition had begun, other illegal, violent actors would more easily 'move in'. An aggravating factor was that, unlike with the illegal economy of illicit crops, where FARC had built mutual economic relationships with communities, with mining it had associated with external mining entrepreneurs who could simply partner with the newcomers to continue. This participant believed the transition away from illegal economies would be difficult to implement, intensifying violence and human rights risks where mining was concerned.

Several specialists saw corruption at the local and national levels as a central enabling factor of illegal mining that remained unaddressed. At the local level, police, army and local authority corruption allowed the free movement of illegal mining machinery from main cities into the most remote areas of the Choco region. At the national level, analysis by the General Controller of Colombia (2012) revealed the gold that paid royalties was much less than the gold that was exported and paid taxes, which to Rudas (Rudas in Ardila, 2015) indicated the possibility of corruption at the Colombian tax agency or in the gold export process. Participants from international agencies expressed concern at this problem. A participant from CREER explained the international community concern thus:

> *The fact of the matter is that not even 30 per cent of the export gold is certifiable as having a legal origin, so this makes other governments and the OECD flinch. That is because there is a value chain and given the circumstances a great proportion of the gold that comes from Colombia has an illicit origin linked to the armed conflict or to organised crime, but is being laundered somehow through institutional channels. For the international community this is a bad state of affairs. Although one part of the Colombian State says this is terrible, another side denies it.*

In parallel, regulatory programs and military solutions to the illegal mining problem were falling short. A participant from CREER, judged the RUCOM – a gold tracking scheme that aims at making the mineral traceable through a unique mineral producer identification – as "*based on good intentions but too far from the local context*", thus ineffective in addressing the pervasive dynamics of illegality and corruption. As an example, this specialist explained, "*with RUCOM the criminal bands demand that miners who are registered with the local government sell them the gold and include an additional amount in the sale declaring it as their own*". Another participant, a mining specialist, found in fieldwork that illegal mercury was gifted to individual miners to build loyalty to a specific buyer. So the traceability initiative fell short in a context of pervasive criminality. Military solutions were ineffective as well, according to this specialist who explained, "*The business is so profitable that burning the machinery and arresting the operator makes no difference, they will just buy another machine and start again*". Most participants believed that a military response was necessary and part of the solution, but that it also needed to be part of a broader strategy. As for the formalisation channels, there were numerous concerns. Most believed that the formalisation policy did not provide sufficient complexity to match the range of mining activities that existed and that it unfairly favoured wealthier miners leaving others in a 'poverty trap'. The poverty trap, discussed the participant from CREER, emerged because "*What is defined as subsistence mining is a very low amount of gold production, and beyond that production the miner is considered illegal*". Neither the traceability program of RUCOM nor the formalisation mechanisms or the military operations were effective responses to illegality.

Most participants insisted that livelihood alternatives should be at the core of Government responses to the problem of illegal mining. However, there was an overall perception of lack of commitment. A participating specialist explained that when discussing shifts from mining to alternative livelihoods, a process known in Colombia as 'reconversion', the Government had a very superficial understanding of the baseline conditions. An example the interviewee provided was that:

> *In a somewhat light way, they tell you that it is cultural matter, that a miner will not change livelihood. But this has changed; it has happened in other places; in Nevada, it has happened, there are no miners anymore. So why would you label this a cultural matter?*
>
> [...]
>
> *That oversimplification has come from the highest levels, at the level of the Vice Ministry. If you look at a recent document called mining policy, there is nothing like that there, the word human rights does not exist, the word sustainability does not exist. It is a superficial and capricious document in the sense that if I name things like I want them to be and I believe it, then they are like that. But it is a bit untruthful because reconversion does not exist inside [....] there is not a single peso allocated to understanding what is reconversion, or how you do it. So in populations with massive numbers of people digging up the rivers [...] if this is not done from the beginning then peacebuilding is going to be very difficult.*

This participant judged that, without well thought-out strategies to help people into alternative livelihoods, illegality would prevail, making it extremely difficult to build peace in the post-agreement period.

Lack of State Presence

All the participants were concerned and frustrated about the lack of State presence in the regions the armed conflict affected the most. The participants believed that only State presence could compete effectively with illegality and that communities should not be left alone to deal with the problems that spring from mining. As FARC retreated, commented a human rights specialist, the concern was that some people were "*moving into the territory to take control of the illicit crops and mining*". So in this scenario, the participant believed it was necessary for the State to control these new forces, support the Life Plans and governance processes of the communities, and revise entry strategies for Indigenous territories. Several participants thought that the Government saw mining as a mechanism for territorial control. There were also questions about post-agreement funding and about international aid models that were contractor heavy and light on local content.

There was an overall perception that the State failed to 'show up' in a range of ways that reinforce the dynamics of illegal mining in the regions. A mining specialist saw an opportunity for the Government to take back the role of purchaser of gold and a need for it to go and "*compete in the regions*" in providing livelihood opportunities for people rather than letting the controllers of illicit economies have free rein. In the absence of this, the participant thought that controllers of illegal mining, having proximity to communities, had contributed to shaping negative attitudes to large-scale mining.

> *How have they been able to build that narrative that says that the enemy is corporate mining? We have seen that and it has been effective, even if there*

> *is no real threat. There is an explanation for it. What happened is that during the Uribe Government this territory was carpeted in mining titles in a corrupt and arbitrary fashion, with complete disregard for the territorial aspirations of the communities that have a mining tradition [...]. So many possibilities were nullified with that, like the possibility to agree on plans with mining communities.*
>
> *Mining formalisation nowadays is limited to formalisation with companies that already have a title. So it is very hard to formalise mining, Indigenous or non-indigenous, when there is no free land. These titles have fed a speculative market that people profit from [...] There is a secondary market. So it becomes a bit of a joke that the subsoil belongs to the State. I can understand that the communities feel something when the private businesses of some people, who are very far away, take precedence over the communities, because people think: this is part of our land.*

A representative from the Mining and Energy Planning Unit spoke of a significant failure on the part of the Government that underpinned the sharp expansion of illegal mining in Colombia, converging with the theme of Government inability to exercise territorial control and effectively compete, providing alternatives to illegal economic activity. A human rights specialist working in the aid sector concurred, stating that the Government had failed to regulate the complexity of mining activity. This participant believed, like several others, that military responses were insufficient and communities could not be left alone to deal with the social complexities of the illegal mining problem, nor left to their own devices in what is clearly an unbalanced interaction with legal or illegal miners. The military solution against illegal mining was also seen as difficult to implement and blind to the social dynamics that lead people into illegal mining.

Participants thought State absence triggered many of the difficulties companies faced. It made companies vulnerable to the conflict and violence in the regions. They were unfairly blamed for long-standing problems that were there before their arrival. The violence of uncontrolled illegal mining was attributed to companies. The central problem was that for investors arriving into regions, including mining investors, the first encounter with communities was one already marred with complaints that sprung from State absence, but that the companies or investors were blamed for.

A Meeting of Perspectives: All Eyes on the Glass Almost Empty

A range of specialist or professional discourses, based on international standards and knowledge, align with the concerns of rural communities. Community members voiced their concerns by speaking from their lived experiences about the consequences of mining on their livelihoods and their rights. Experts from human rights organisations arrived at similar, worrying conclusions through analyses based on field data, international norms such as the ILO 169, the UNDRIP, the United Nations Principles on Business and Human Rights, the jurisprudence of the Inter American Human Rights Court, and globally adopted concepts such as the notion of Sustainable Development (Brundtland, 1987). Some industry and Government representatives referred not only to the rights literature and standards but also to the social responsibility discourse within the mining industry and to the extensive literature on the social, environmental and rights consequences of mining in rural communities, again, to land at very similar conclusions. Close to two decades of evolving industry standards (for example ICMM, 2003, 2009, 2011, 2012, 2015), and widely publicised research about the unintended consequences of mining (Ballard & Banks, 2003; Hilson, 2002; IIED & WBCSD, 2002) inform their discourse. In recent years, there have been research exercises about mining undertaken in Colombia that speak to questions of impacts, governance and participation, conflict and armed conflict, and rights (Arbeláez-Ruiz, 2015; CINEP, 2012; Colombian Ombudsman, 2010; de Angulo et al., 2016; Garay, 2013a, 2013b, 2014; Moor & Sandt, 2014; Negrete, 2013; Pardo, 2013; Pardo, Rudas, & Roa, 2014; Pérez Rincón, 2014; Rettberg & Ortiz-Riomalo, 2016; Weitzner, 2012). Building on that body of knowledge, participating specialists' concerns also coincide with the concerns of rural community participants.

So why is there such a confluence of perspectives? It can be argued that the baseline conditions, related to State absence, are precarious. In the Colombian context, industry members are seen as overly keen to start operations, but even for some industry participants, the arrival of the mining industry to some regions of Colombia was not advisable within the conditions of the time. There were shared concerns about an absence of the basic guarantees that would allow the healthy arrival of a mining operator. In such a high-risk environment, a post-agreement transition, where the State has been absent on account of the armed conflict and where corruption, violence and criminality prevail, there was overall apprehension about possible trajectories for mining in the post-agreement phase. Mining could exacerbate or create problems relevant to the diverse professional audiences analysing mining.

Despite this, participants thought the Government would seek to drive growth in mining without resolving the most significant problems. An academic specialising in environment believed the Government would promote large-scale mining investment in an urge to fill the budgetary hole resulting from decreased oil prices. Furthermore, this academic thought it was likely that as communities resisted new mining projects, new pejorative labels would emerge from the Government apparatus to vilify social protest so that resistance towards mining would be repressed in the regions.

In addition, there were no signs that post-agreement mining governance would address the existing problems. Studying the conditions that support positive outcomes from mineral resource extraction in post-conflict scenarios, Nichols, Lujala and Bruch (2011) have argued that natural resource governance is a space of significant risk and opportunity for post-conflict peacebuilding, because natural resources can trigger, prolong and reignite conflict. "*By restoring capacity to govern natural resources, addressing legal pluralism, effectively managing decentralization, introducing adaptive governance, and ensuring transparency and accountability, post-conflict countries can make significant strides in putting in place more effective and durable governance structures*", said the authors (Nichols, Lujala & Bruch, 2011, p. 23). While it is not my purpose here to evaluate Colombia's mineral resource governance capacity, the participants cited situations, besides corruption and violence, that indicated gaps on the fronts these authors suggest: there was a large illegal gold mining industry with significant amounts of that gold exiting through the legal export avenue (Ardila, 2015; General Controller of Colombia, 2012); the question of subsoil ownership remained contentious, demonstrating a clash between Indigenous law and mainstream law; and local popular consultations to block Government-backed mining projects proliferated in the five years leading up to 2018, suggesting frictions rather than effective decentralisation in mining-related decisions. As no change in mining governance was anticipated for the post-agreement transition, the conditions were unlikely to improve. But there were diverse readings of the situation and, for that reason, there were divergences that I discuss in the next section, where I cite a different interpretation of the natural resource governance discourse in the Colombian context.

Divergences

The clashing perspectives become visible when contrasting the previous views with those of Colombian Government mining institutions convinced of the convenience of industrial mining expansion through foreign direct investment. Interviews with Government mining institution officials indicated that the National Government had not signalled willingness to change the management of mining in Colombia in the post-agreement transition, neither through the Peace Agreement nor in its implementation. The signal that National Mining Agency participants were receiving was that the Government would concentrate on two matters. The first was the formalisation of rural areas including mining within the existing parameters. The second was seeking a pathway for legislation changes to respond to the Constitutional Court requirement that mining project development be negotiated between the National Government and territorial entities. There were pockets of interest and more elaborate understandings of the interaction between mining, illegality and armed conflict together with efforts to design strategies for the mining sector to enter into dialogue with territories, specifically at the Mining and Energy Planning Unit. However, it was unclear whether these approaches had traction with other Government mining institutions. The overall strategy was to forge ahead, in a business-as-usual fashion.

The public narrative of institutions, such as the National Mining Agency and the Ministry of Mines and Energy, concentrated on the role of mining as a Government revenue source and a lever to address poverty in the regions and bring 'development'. The targets include increasing gold and coal production and increasing the number of formalised mining ventures by 2,000 (Ministry of Mines and Energy, 2018a, 2018c). Senior officials argued causal links between royalty income, mining taxes and poverty reduction. For example, the Minister for Mines and Energy stated, "*Hydrocarbons and mining will generate royalties, taxes, and dividends to pay for energy, roads, education, health and a better Colombia*!" (Suárez, 2018a) and that royalties explained 20 per cent of poverty reduction in Colombia (Suárez, 2018b). The Minister added that "*as far as mining goes, we are going to work on formalisation, we are going to attack illegal mining and formalise the small miners*" (Ministry of Mines and Energy, 2018b). Strategies granted limited attention to the 'poverty trap' in formalisation, or to the human rights risks that legal mining represents, or to the cultural opposition to mining in some communities, not to mention Colombia's high levels of corruption. The narratives I heard in interviews mirrored this approach. The experience of officials from the National Mining Agency led them to believe that illegal mining activity created a negative reputation for large-scale mining that made its entry difficult. But they believed it was precisely that latter kind of operator that was best qualified to conduct mining given the contractual, financial, environmental and social obligations that the Colombian law sets out. Like other participants warned, there was no visible acknowledgement of the human rights risks attached to mining, legal or illegal, merely a focus on the convenience of large-scale mining operators. Furthermore, these Government officials found that interactions with ethnic communities were difficult because the Agency is an executor, not a policy-maker or a legislator, and that it would be very difficult for the law to respond to the range of mining scenarios to afford the differential treatment that some ethnic communities required. Others have interpreted this type of positioning as a one-size-fits-all mining policy (de Angulo et al., 2016; Viana, 2015).

Although participants from rural communities, multilateral and observer organisations, aid organisations, as well as experts and academics, and some from the State saw legal and illegal mining as having clear links with the armed conflict and violence, there were some participants from industry and Government that saw legal mining as divorced from these dynamics. A participant from the State's Attorney described the results of extensive investigations on large-scale illegal mining operations that were ultimately connected to what has been termed "*amphibious mining companies*". These are legal mining companies "*that pay tax and have flashy offices in Medellin*" said the participant, and have legal mining sites to which they attribute the production from environmentally and socially damaging large-scale illegal operations they finance in Cauca and elsewhere. Operations to bring down these criminal syndicates have involved the coordination of several State bodies and thousands of State forces officers (army, navy and air force). That the link between some legal mining operators and illegality is not acknowledged in some sectors of Government and industry makes for a stark contrast of perspectives.

The disparate views on the relationship between mining, illegality and armed conflict result in contrasting opinions about the trajectory or role of mining in the post-agreement period. Some participants from industry concur with the Government view of mining as a contributor to the 'formalisation of rural areas' and to development. Some participants were keen on seeing uncertainties over access to land and community participation resolved. One of them stated they were "*losing invested money*". Another industry participant, however, aware of the sensitive post-agreement reality and existing conflicts about land, thought it was not appropriate to enter as a new economic actor to add to the pressures in some regions. In other regions, this participant thought what was necessary was to look beyond the transactional matter of access and think about how industry can leverage State presence, how it can support more community participation in decisions related to mining, how it can best direct its social investment and how it can contribute to 'territorial consolidation' – understood as strengthening the local community organisations and leaderships that are going to be industry's liaison points and partners in promoting better outcomes for host communities. Even within the industry itself, the perspectives on industry's role in the transition period varied widely.

Despite the overwhelming message from most interviewees, including some from industry, being that there is not enough rural community participation and specifically Indigenous participation in decision-making about mining, there remained some legalistic interpretations that saw the separation of soil and subsoil as an enabling platform for industry entry into territories and a basis for more limited general and differentiated community participation. Beyond the group of interviewees, during the fieldwork I came across members of industry who thought there was too much participation in Colombia, because of the number of prior consultations implemented. So the ground was fertile for ongoing contestation and likely social and environmental conflict surrounding mining.

The Glass Half Full: Good News from the Mineral Resource Governance Camp?

Earlier on, I took issue with the State building focus (Beevers, 2010; Unruh, 2014) and disregard for active community roles (Beevers, 2015) characteristic of the post-conflict mineral resource governance literature that provide formulas to manage mineral wealth to maximise the positive effects of mineral resource extraction and its contribution to peace. This well-intentioned literature and professional discourse is nevertheless of significant value in bringing together international experience where the role of the State is concerned. However, when not informed with rights, livelihoods, environmental, social, and conflict and peace dynamics analyses specific to the context, it can distance itself from what is happening in mining governance on the ground, and overemphasise elements that relate only to the positives and resonate with the needs of Governments, resulting in an overly optimistic discourse about revenues and economic growth. Let us take some examples from the tremendously influential work of the Natural Resource Governance Institute (NRGI), formerly Revenue Watch, creators of the Resource Governance Charter, the blueprint for mineral resource governance. The Charter requires a set of local conditions for good resource governance including, through its Precept One, that countries define whether to open areas for exploration and define an inclusive and comprehensive national strategy for the benefit of all citizens. Most would argue this would be of great benefit to Colombia. The problematic aspects emerge when these generalised prescriptions are applied or evaluated in an aggregate fashion and with insufficient awareness of the local context. As illustration of what was an incomplete understanding of the national context, the NRGI issued a stigmatising statement, in 2015, in the opening of its strategic note on Colombia:

> *The Colombian constitution of 1991 was approved at a time when Colombia was not thinking about oil and mining as leading sectors for economic growth and sustained development. It established strong provisions regarding the environment and territorial and consultation rights for Indigenous and Afro-Colombian peoples. These provisions later proved to hamper mining and hydrocarbons from becoming primary drivers of growth and prosperity.*
>
> (NRGI, 2015, p. 1)

Similar statements, that portray Indigenous and Afro-Colombian people as one of the obstacles to extractive activities and prosperity, have emerged from powerful media outlets and sectors of the Colombian Government before (Orduz, 2014)[3]. In a polarised environment, characterised by violence against social leaders that oppose development projects, such claims become very dangerous, so these domestic statements caused significant concern in the social sector at the time. In the domestic environment, these are part of a contest for influence over decisions about land use and resource extraction. The efforts at stigmatising Colombia's ethnic populations are well documented and sufficiently damaging. But, taken out of context and validated in a credible international platform, these claims become a far more serious problem. They present mining as a safe path to prosperity and collective rights as obstacles, rather than necessary conditions, and configurative forces, for sustainable development models. In a blanket statement, all extractive activity leads to prosperity and all prior consultation "*hampers*" development. Add the NRGI brand and such understandings can continue to gather legitimacy in some sectors and contribute to further polarisation. We observe here a less positive side of the resource governance discourse.

However, NRGI has also partnered with Foro Nacional por Colombia (Foro) – a reputable Colombian NGO with strong research capabilities – and has, through that collaboration, issued more complex and critical analyses that distance themselves from the blunt 2015 statement on p. 198. Foro's subsequent report concluded that

> *[a]lthough the national government states that the extractive sector generates important growth in public investment in strategic areas for social development promotion and increased territorial convergence, people in Colombia are left with the sensation that extractive revenues have not contributed, as expected, to the country's development.*
>
> (Peña Niño & Martínez, 2016, p. 45)

Foro's analysis of living standards in municipalities and departments indicates that it is those with important extractive production shares that display stronger indices of poverty, exclusion and inequality. That structural problems remained unresolved during the mining boom of the 2000s led Foro to emphasise that without political will, economic growth alone is insufficient to tackle social problems. Moreover, Foro warned of the complex path ahead as Colombia faced the upcoming costs of Peace Agreement implementation amid growth deceleration (Peña Niño & Martínez, 2016). The Government statements illustrate that its discourse does not reflect an awareness of these findings, neither were they widely discussed in the media.

In contrast, NRGI's 2017 ranking of Colombia as the 10th out of 70 countries evaluated on their mineral resource governance (NRGI, 2017) was widely publicised. What was not so specifically reported was that the ranking indicates NRGI judges Colombia's mining governance as "*satisfactory*" rather than "*good*", and that the ranking is an aggregate instrument that compiles a range of parameters most of which refer to legal activities, so it is hard to reconcile with the tangible consequences of a large illegal mining sector. The index is based on a 150-question questionnaire (see NRGI, 2017) that cannot assess the granularity of specific Colombian local contexts. The NRGI (2017) did emphasise the lack of legal enforcement and the high incidence of illegality or informality, as well as concerns about social and environmental impacts and the participation of local communities. Nevertheless, the ranking itself was received as a positive story in Colombia. It is this kind of governance expectation or prescription framework that informs Government mining institutions' aspirations in Bogotá, where efficiency discourses prevail and measurement is crucial. So what better than a ranking against specific governance parameters?

Evident in the focus on formal industry growth – through formalisation and foreign direct investment, without a critical perspective – is a partiality towards mining that concentrates solely on its positive potential. Beevers (2019) documented this tendency among international peacebuilders with regards to natural resource exploitation in post-conflict Sierra Leone and Liberia. Anthropologist David Trigger described a "*culture of development ideology*" among mining professionals in some Australian mining sites, who saw their work as a force for good, and understood geologic discoveries not only as landmarks for their professional careers but also as promising milestones towards a bright future for the locals (Trigger, 1997). Former ICMM CEO, Anthony Hodge (2018) has suggested that the emergence of this optimistic view of mining among industry professionals relates to the post-second world war period, where societies concentrated on creating new economic opportunity and promoted well-intentioned efforts towards industry expansion at a time when the knowledge of today was not available. That knowledge of today has called for new ways of thinking, but existing patterns appear to be well entrenched.

Possibilities and Limitations

The convergences and distancing points in the views of participants in this chapter outline a domain of limitations but also of possibilities for mining in Colombia. For most participants, the role of the Government was central to these possibilities and limitations. The participants described a situation where the Government was unable or unwilling to cope with the complexity of the mining sector, had poorly set up relationships with regional actors, fuelled a polarised debate and failed to implement appropriate community participation. In 2014, the Colombian State itself had diagnosed limited institutional capacity in the young Government bodies of the sector (Morelli, 2014). The question is whether capacity was evolving. Nevertheless, there appeared to be a misguided understanding of State capacity where the central Government institutions either assumed the regional capability was there to respond to the rapid transformations and risks that ensue with mining, or had short-sighted understandings of the social and environmental risks of the activity. Furthermore, the pressing questions about rural livelihoods that underpin some of the informal or illegal mining entrepreneurship remained unaddressed in Government mining policy, where the weight of the efforts went into a formalisation approach that some have called a 'poverty trap'. The collapse of the oil price drove a fiscal deficit and a Government urge to increase revenues. So blunter instruments and elements of the natural resource governance discourse that favour the idea of mining industry growth are emphasised, whereas more complex analyses, including those coming from the same area and calling for equally complex responses (see Peña Niño & Martínez, 2016), receive limited attention and become less influential in Government institutional discourses.

Although it has been argued that "*reforming governance of natural resources and their revenues is one of the most important measures for peacebuilding in post-conflict countries*" (Nichols, Lujala & Bruch, 2011), there was no political will to include reform of the extractive industry governance as part of the Peace Agreement. Nevertheless, some participants were able to identify avenues where possibilities existed to influence change in mining dynamics. According to these participants, whether the concerns about mining-illegality-conflict interactions could be resolved would depend to a great extent on whether communities could make their concerns heard or act to resolve them. A participant with an in-depth understanding of the Peace Process believed that if Colombian mining policy were to follow the principles of the Peace Agreement and prioritise understanding community preferences and concerns, like was done in the case of the drug economy, this would likely cause a positive shift. However, the prospects of that materialising were highly uncertain. What was more tangible, in terms of communities being able to seek resolution for problems with mining in the post-agreement transition, was the established participation mechanisms and the emerging participation spaces attached to the Peace Agreement implementation. Another participant, a peace researcher, considered the limited mining content in the Peace Agreement a gap, but identified opportunities to advocate for Indigenous rights in the face of mining through elements such as the holistic rural reform, including its emphasis on small producers; the provisions for victims and political participation; the priority that regions affected by conflict and poverty receive; and the gender approach. In-depth knowledge of these Agreement provisions would be fundamental for the communities to protect their rights in the context of mining. However, this peace researcher highlighted that much would depend on the balance of power. Considering the Government approach was to cordon off any discussion of the economic model, this researcher believed Indigenous communities would be likely to obtain better results if, when discussing mining concerns, they tailored the language to what was explicit in the Agreement rather than concentrating discussions around the economic model. In this way, the communities could talk about the environment, about gendered impacts, about consequences on young people, about effects on Indigenous rights, etc., all of which are included in the Agreement and are causes the international community backs.

Nasa organisations – seeking response to the problems caused by mining, and protections for traditional mining – can build on the echoing perspectives of several specialist audiences, such as the powerful multilateral and observer actors, to promote a more nuanced reading in Government institutions of the mining dynamics in Indigenous territories. There were broadly consistent readings between Nasa communities and some people from industry or even the Government who could – despite the overall Nasa apprehension towards mining companies and Government mining institutions – become unlikely, but fruitful alliances. For example, there were participants from industry who, aware of their capacity to leverage State presence, were cognisant of the need to understand how to best influence such a shift for the benefit of communities. Furthermore, industry participants were interested in more social dialogue, and the Ministry for Mines and Energy, at least in public discourse, was by 2018 promoting the merits of dialogue as a component of its strategy in the regions. These possibilities, together with the Agreement mechanisms, might provide some opportunities. But once again, it is the communities who are called on to be the protagonists.

Notes

1 An aggravating factor is the privatisation trend in mining security in Colombia and Latin America, where States often do not regulate private military and security service providers (Perret, 2013).

2 This coincides with the diagnoses of Colombian free, prior and informed consultation and consent scholars (Viana, 2016).

3 Orduz (2014, p. 24) discussed how Colombia's most read magazine, *Semana*, whose director was a first cousin of President Santos, referred to prior consultation as "*the stick in the wheel of development*"; and major newspaper, *El Tiempo*, concurred, citing the then Minister of Agriculture as stating that prior consultation had turned into an "*interminable and excessively costly labyrinth that had reached a crisis point*".

References

Arbeláez-Ruiz, D. (2015). *La Inclusión Social en las Políticas y Planes Públicos para el Sistema Minero de Colombia: reflexiones y aprendizajes.* Brisbane, Australia: Centre for Social Responsibility in Mining, The University of Queensland. Retrieved from www.csrm.uq.edu.au/media/docs/1286/csrmaprendizajes.pdf.

Ardila, N. (2015). Oro ilegal en Colombia, depredador de las finanzas y el ambiente. *Catorce6 Revista Ambiental.* Retrieved from www.catorce6.com/investigacion/350-publicaciones/13467-oro-ilegal-en-colombia-depredador-de-las-finanzas-y-el-ambiente

Ballard, C., & Banks, G. (2003). Resource wars: The anthropology of mining. *Annual Review of Anthropology,* 32, 287–313.

Beevers, M. (2010). *How does natural resources governance shape postconflict peacebuilding? Assessing problems and opportunities.* APSA 2010 Annual Meeting Paper. Available at SSRN: https://ssrn.com/abstract=1642152.

Beevers, M. (2015). Governing natural resources for peace: Lessons from Liberia and Sierra Leone. *Global Governance: A Review of Multilateralism and International Organizations,* 21(2), 227–246. doi:10.5555/1075-2846-21.2.227.

Beevers, M. (2019). *Peacebuilding and natural resource governance after armed conflict: Sierra Leone and Liberia.* Cham, Switzerland: Springer.

Brundtland, G. H. (1987). *Our common future.* Oxford: Oxford University Press.

CINEP. (2012). *Minería, conflictos sociales y violación de Derechos Humanos en Colombia.* Bogotá, Colombia: Centro de Investigación y Educación Popular / Programa por la Paz.

Colombian Ombudsman. (2010). *Minería de Hecho en Colombia: Defensoría delegada para los Derechos Colectivos y del Ambiente.* Bogotá, Colombia: Colombian Ombudsman. Retrieved from www2.congreso.gob.pe/sicr/cendocbib/con4_uibd.nsf/F11B784C597AC0F005257A310058CA31/%24FILE/La-miner%C3%ADa-de-hecho-en-Colombia.pdf

de Angulo, L. F., Miranda, C., Llorente, B., Tobón, S., Torres Condía, N., Castillo Méndez, A., … Arango, V. (2016). *Evaluación Integral Sectorial de Impactos en Derechos Humanos - La minería que no se ve.* Bogotá, Colombia: Centro Regional de Empresas y Emprendimientos Responsables.

Garay, L. J. (2013a). Introducción. In L. J. Garay (Ed.), *Minería en Colombia: derechos, políticas públicas y gobernanza.* Bogotá, Colombia: Contraloría General de la Nación, Colombia.

Garay, L. J. (Ed.). (2013b). *Minería en Colombia: derechos, políticas públicas y gobernanza*. Bogotá, Colombia: Contraloría General de la Nación, Colombia.

Garay, L. J. (Ed.). (2014). *Minería en Colombia: Control público, memoria y justicia socio-ecológical, movimientos sociales y posconflicto*. Bogotá, Colombia: Contraloría General de la Nación, Colombia.

General Controller of Colombia. (2012). *Informe. del Estado de los Recursos Naturales y del Ambiente 2011–2012*. Bogotá, Colombia: General Controller of Colombia.

Hilson, G. (2002). An overview of land use conflicts in mining communities. *Land Use Policy*, 19(1), 65–73. doi:10.1016/S0264-8377(01)00043-6.

Hodge, A. (2018). *Three months of adventure at SMI*. Presented as part of the SMI Seminar Series. December 6. Sustainable Minerals Institute, Brisbane, Australia.

International Council on Mines and Metals (ICMM). (2003). *Sustainable development framework: 10 principles*. Retrieved February, 2019, www.iucn.org/sites/dev/files/import/downloads/minicmmstat.pdf. London: ICMM.

International Council on Mines and Metals (ICMM). (2009). *Human rights in the mining and metals industry: Overview, management approach and issues*. London: ICMM.

International Council on Mines and Metals (ICMM). (2011). *Voluntary principles on security and human rights: Implementation guidance tool (IGT)*. London: ICMM.

International Council on Mines and Metals (ICMM). (2012). *Human rights in the mining and metals industry: Integrating human rights due diligence into corporate risk management processes*. London: ICMM.

International Council on Mines and Metals (ICMM). (2015). *10 Principles*. London: ICMM.

International Institute for Environment and Development (IIED), & World Business Council for Sustainable Development (WBCSD). (2002). *Breaking new ground: Mining, minerals and sustainable development* (International Institute for Environment and Development & World Business Council for Sustainable Development, Trans.). Sterling: Earthscan Publications.

Ministry of Mines and Energy. (2018a). (@MinMinas) *"En cuanto a minería vamos a aumentar la producción de oro y de carbón"*: @mafsul #ElFuturoYaComenzó. 17 November, Tweet.

Ministry of Mines and Energy. (2018b). (@MinMinas) *"En materia de minería, vamos a trabajar en la formalización, vamos a atacar la minería ilegal y formalizar a los pequeños mineros"*: @mafsul #ElFuturoYaComenzó. 17 November, Tweet.

Ministry of Mines and Energy. (2018c). (@MinMinas) *"La meta será en formalización minera, 2000 unidades que equivalen a 10 mil personas que van a entrar en la legalidad"*: @mafsul #ElFuturoYaComenzó. 17 November, Tweet.

Molano, A. (2015). Fragmentos de la Historia del Conflicto Armado (1920–2010). In Comisión Histórica del Conflicto y sus Víctimas (Ed.), *Contribución al Entendimiento del Conflicto Armado en Colombia*. La Habana: Comisión Histórica del Conflicto y sus Víctimas. Retrieved from https://indepaz.org.co/wp-content/uploads/2015/02/Version-final-informes-CHCV.pdf

Moor, M., & Sandt, J. v. d. (2014). *El Lado Oscuro del Carbón: violencia paramilitar en la zona minera del Cesar*. Colombia. Utrecht, Netherlands: PAX Paises Bajos.

Morelli, S. (2014). Prólogo. In L. J. Garay (Ed.), *Minería en Colombia: Control público, memoria y justicia socio-ecológical, movimientos sociales y posconflicto*. Bogotá, Colombia: Contraloría General de la Nación, Colombia. Retrieved from https://redjusticiaambientalcolombia.files.wordpress.com/2014/08/libro-mineria_contraloria_vol-iv.pdf

Negrete, R. (2013). Derechos, minería y conflictos. Aspectos normativos. In L. J. Garay (Ed.), *Minería en Colombia: Derechos, políticas públicas y gobernanza.* Bogotá, Colombia: Contraloría General de la Nación, Colombia. Retrieved from https://redjusticiaambientalcolombia.files.wordpress.com/2013/12/libro_mineria_contraloria-2013.pdf

Nichols, S. S., Lujala, P., & Bruch, C. (2011). When peacebuilding meets the plan: Natural resource governance and post-conflict recovery. *The Whitehead Journal of Diplomacy and International Relations*, 12(1), 11–26. Retrieved from www.proquest.com/docview/875634968/fulltextPDF/50F7357403924292PQ/1?accountid=14723

NRGI. (2015). *Colombia Country strategy note* (p. 1). NRGI. Retrieved from https://resourcegovernance.org/sites/default/files/documents/nrgi_colombia-strategy_20160629.pdf

NRGI. (2017). *Colombia ranked in world's top ten for oversight of natural resources despite concern over management of revenues.* Retrieved from https://resourcegovernance.org/news/colombia-ranked-world-top-ten-oversight-natural-resources-despite-concern-over-management.

Orduz, N. (2014). La Consulta Previa en Colombia. *Documentos de Trabajo ICSO. Santiago de Chile: Instituto de Investigación en Ciencias Sociales UDP.* N° 3. Retrieved from https://icso.udp.cl/publicacion/la-consulta-previa-en-colombia/

Pardo, A. (2013). Propuestas para recuperar la gobernanza del sector minero colombiano. In L. J. Garay (Ed.), *Minería en Colombia: derechos, políticas públicas y gobernanza.* Bogotá, Colombia: Contraloría General de la Nación, Colombia. Retrieved from https://redjusticiaambientalcolombia.files.wordpress.com/2013/12/libro_mineria_contraloria-2013.pdf

Pardo, Á., Rudas, G., & Roa, E. (2014). Análisis de los contratos de Drummond Ltd. en el departamento del Cesar. In L. J. Garay (Ed.), *Minería en Colombia: control público, memoria y justicia socio-ecológical, movimientos sociales y posconflicto.* Bogotá, Colombia: Contraloría General de la Nación, Colombia. Retrieved from https://redjusticiaambientalcolombia.files.wordpress.com/2014/08/libro-mineria_contraloria_vol-iv.pdf

Peña Niño, J., & Martínez, M. F. (2016). *Los Efectos del BOOM de las INDUSTRIAS EXTRACTIVAS en los Indicadores Sociales.* Lima, Perú: Foro Nacional por Colombia.

Perret, A. (2013). Las compañías militares y de seguridad privada en el sector minero-energético: un desafìo para la aplicación de los derechos humanos. In A. C. G. Espinosa (Ed.), *Los retos de la gobernanza minero-energética* (Vol. 43). U. Externado de Colombia.

Pérez Rincón, M. (2014). Conflictos ambientales en Colombia: Inventario, caracterización y análisis. In L. J. Garay (Ed.), *Minería en Colombia: control público, memoria y justicia socio-ecológical, movimientos sociales y posconflicto.* Bogotá, Colombia: Contraloría General de la Nación, Colombia. Retrieved from https://redjusticiaambientalcolombia.files.wordpress.com/2014/08/libro-mineria_contraloria_vol-iv.pdf

Rettberg, A., & Ortiz-Riomalo, J. F. (2016). Golden opportunity, or a new twist on the resource-conflict relationship: Links between the drug trade and illegal gold mining in Colombia. *World Development*, 84, 82–96. doi:10.1016/j.worlddev.2016.03.020.

Suárez, M. F. (2018a). (@mafsul) "*#ElFuturoYaComenzo Trabajaremos por EQUIDAD para generar oportunidades para todos los Colombianos. Hidrocarburos y Mineria generaran regalias, impuestos y dividendos para pagar energia, vias,*

educacion, salud y tuna (sic) mejor Colombia !jorgemesastman !VictoriaBxE !MinSaludCol." 17 November, Tweet.
Suárez, M. F. (2018b). (@mafsul) "*EQUIDAD – contribuiremos con el presupuesto de regalías mas alto de la historia, éstas generan progreso en territorio – explican el 20% de la disminución de la pobreza en Colombia*" #ElFuturoYaComenzó. 17 November, Tweet.
Trigger, D. (1997). Mining, landscape and the culture of development ideology in Australia. *Ecumene*, 4, 161–180.
Unruh, J. (2014). Multi-sector capacity needs in challenging the resource curse in conflict-affected countries. *International Journal of Peace Studies*, 19, 2.
Verdad Abierta. (2015). *La petroguerra del ELN en Arauca Verdad Abierta*. Retrieved from https://verdadabierta.com/la-petro-guerra-del-eln-en-arauca/.
Viana, A. (2015). *Mapa de Un Sistema Minero Incompleto. Informe de análisis de contexto preparado para el proyecto Promoviendo la Inclusión Social en la Planeación Minera Nacional*. Brisbane, Australia: Centre for Social Responsibility in Mining, The University of Queensland. Retrieved from www.csrm.uq.edu.au/publications/mapa-de-un-sistema-minero-incompleto
Viana, A. (2016). *El derecho a la consulta previa - echando un pulso a la nación homogénea*. Bogotá, Colombia: Editorial Pontificia Universidad Javeriana.
Weitzner, V. (2012). *Executive summary: Holding extractive companies to account in Colombia — An evaluation of CSR instruments through the lens of Indigenous and Afro-descendent rights*. Bogotá, Colombia: The North-South Institute, Proceso de Comunidades Negras, Resguardo Indígena Cañamomo Lomaprieta.

9 Conclusion

The Nasa of North Cauca's experience with mining and their moral principles have shaped a predominant view that most mining leads to moral breaches and puts the aspirations and dreams of the Nasa people at risk. In light of Nasa cosmogony, interventions of, and engagements with, the natural world carry moral consequences. Within the Nasa ontology, entities in the natural world have agency and moral and emotional propensities and capacities. External actors are not considered appropriate to conduct mining for fear of irreversible degradation of the Nasa world. Gold mining is seen as particularly dangerous. Gold is construed as a powerful entity whose energy can only be harnessed by ritual experts. By 'touching gold' without the necessary ritual resources, the Nasa risk being compelled into morally damaging behaviours that put the individual ahead of the community in a tireless search for more and more gold. As I illustrated in Chapters 3 and 6, a cycle of degradation begins with the miner and filters into the miner's community and natural environment. This can ultimately precipitate retribution from Mother Earth, should the Nasa fail to fulfil their reciprocity obligations.

In Chapter 3, I demonstrated that the Nasa saw mining as morally flawed based on the ideal of Living Well, moral principles, and political, economic, cultural, spiritual and ecological aspirations. Living Well is living in harmony with the self, with others and with Mother Earth. It requires honouring an integrative relationship with this motherly being and with the community. It is contingent upon a dignified life and robust economic activity. To most participants, unless the Nasa conduct mining in an artisanal way in harmony with the mandates of Nasa Authorities, mining unleashes forces that clash with the idea of Living Well and with the principles of Unity, Land, Culture and Autonomy – the four moral pillars of the Nasa discourse. Mining that deviates from these parameters divides communities between supporters and detractors and destroys unity. It degrades land as the basis for agriculture; it degrades the territory that provides a space to recreate identity and memory, to heal and to realise and protect collective rights; and it degrades the motherly being that the Nasa are part of and entirely dependent upon. With uncontrolled mining, new wealth and patterns of consumption arise, so the Nasa culture, and the system that reproduces it, weaken. Therefore, the

DOI: 10.4324/9781003226895-9

possibilities of a response to mining consistent with Nasa knowledge about the world decline. This kind of mining creates power shifts because, through cash and weapons, external actors exert pressures that diminish the Nasa's capacity to self-govern according to their dreams and aspirations. Nasa autonomy declines. This moral discourse speaks of a kind of mining that is environmentally damaging, violent and expansive; that displaces established relationships and networks of reciprocity; and that does not build on the Nasa understanding of the natural world or on Nasa forms of government. To most participants this industrial or unauthorised mining impedes the harmony and freedom that Living Well encapsulates, and contravenes the moral principles that embed and support Nasa organisations' aspirations and dreams.

Despite this predominant view, there were alternative perspectives because some Nasa communities and individuals also practice mining. These cases have invited other understandings of mining, as I discussed in Chapter 4. Some Nasa territories found that only by practicing mining themselves would the existing legal framework and geopolitical environment allow them to keep external mining entrepreneurs at bay. In other cases, precarious material conditions, lack of economic options and individual initiative saw the emergence of Nasa mining entrepreneurs who trespassed the mandates of their Authorities, engaging in a dangerous but personally rewarding relationship with gold. These miners became the subject of internal social and political dissent and were at the centre of local conflicts about mining. Still others, who did not practice mining, wanted to see more analysis and more empathetic responses to the plight of the miners rather than a blanket NO. The NO was, to some, a political position that needed to evolve to become more complex, based on hard evidence and on dialogue between proponents and opponents.

The Nasa miners, sanctioned or not, were trespassing and faced tensions they attempted to resolve. In this effort they invoked ideas of fairness, created new alliances, prioritised some moral values over others or compartmentalised their moral understanding of mining. As I illustrated in Chapter 4, a Nasa Authority might decide to practice mining rather than to allow into its territory a mining company they do not trust and who they believe will extract the benefits and leave to the community little more than social and environmental damage. A miner might consider the rewards of mining fairer than those local agricultural and job markets can offer. This miner might create stronger alliances with fellow miners. Or a Nasa territory that has decided to do mining might seek to enhance relationships with neighbouring mining communities with mining traditions. A Nasa miner prioritises the duty to look after family or the idea of fair economic reward over the duty to care for Mother Earth within Nasa parameters. A Nasa territory with its own mine might put autonomy in the face of external mining proposals first, over unity in a consolidated position of NO to mining with other Nasa territories. Nasa Authorities might compartmentalise and decide it is correct for them to practice mining, but not for a company that has no understanding of the

Nasa ontology, ecology and culture. Individual miners can likewise compartmentalise and state that it is acceptable for them to practice mining in the present, but not for their children to do it when they grow up. Within these difficult contexts and tensions, the moral choices and understandings of the miners are complex, and resist simplification and generalisation.

Nevertheless, the predominant response was one of opposition to mining and this governed a multilayered and specifically territorialised resistance to mining in Nasa territories that I analysed in Chapters 5 and 6. The response included a spiritual element setting the course and principles for action; a policy and legal element providing an infrastructure of norms and mandates; an institutional design, including an economic-environmental system to facilitate alternative livelihood projects; research, education, dialogue and advocacy; as well as territorial control or enforcement at the hands of the Guardia, who both monitored the situation with mining and executed the orders of the Authorities. The responses in the organisational domain of ACIN were consistent with those I observed in the Nasa Resguardo (territory) of Canoas because their leadership converged in a position of NO to mining. However, the response could vary because the Nasa organise themselves in highly localised autonomous Authorities that are the highest level authorities. Just as Canoas converged with ACIN in responding to mining, Resguardos like Delicias, who own a gold mine, have experienced tensions with ACIN on mining. Resistance to mining was predominant, but the principle of autonomy allowed for significant difference across territories.

Nasa ideas of peace and peacebuilding, and Nasa resistance work and identity, shaped Nasa responses to mining in the conflict-to-peace transition context. The Nasa have diverse understandings of peace that converge in their alignment with notions of positive peace and Living Well. As I demonstrated in Chapter 7, the participants understood peace as not only negative peace, or the absence of violence and the end to the armed confrontation, but also as positive peace or the idea of realising rights, freedoms and equity that in turn resonates with the concept of Living Well, in harmony. Where the aspiration of Living Well is realised there are no factors causing a resurgence of violent confrontation, because a dignified life is accessible, and the Nasa can reach their dreams and aspirations as a community.

But the reality had strayed far from the ideal of negative peace, let alone positive peace or Living Well, and the Nasa resisted mining as an economic activity that they connected to violence and the armed conflict (see Chapter 7). The Nasa remained in their territories amid the armed conflict. Resistance became embedded in the design and work programs of Nasa organisations and in the configuration of social mobilisation, and it inspired some of the most emblematic elements of the Nasa Indigenous organisations such as its Indigenous Guardia, winner of the National Peace Award. In parallel, mining consolidated as a key financing activity for illegal armed actors, who are particularly active in Cauca and have taken the lives of numerous leaders during the post-agreement transition. The Nasa experience with the armed conflict and the drug trade that financed it has made them, by force of need, reluctant but skilled analysts of the dynamics of violence and armed conflict. The Nasa ably discern the symptoms of violent illegal economies and have drawn strong parallels between the violent and deteriorating symptoms of the drug trade and the illicit mining economies. Furthermore, many of the Nasa participants saw mining, legal or illegal, as a continuation of the processes of violence, oppression and dispossession that began with the arrival of the Spanish to the Americas, and legal mining as a magnet that attracts violent illegal miners and violent State repression. As a result, there was consensus that resisting external mining entry, legal or illegal, was a peacebuilding effort. The resistance to mining, therefore, followed similar contours to the resistance to the armed conflict and to violence: a resistance that constitutes a building block of Nasa identity.

There was less confluence, however, on whether banning mining by Indigenous people can contribute to building peace. In this regard, the tensions remained. Nasa understandings of peace are strongly linked to a dignified existence in the material sense. The question of livelihoods is central to the notion of peace. There was contention and conflict around mining by Nasa community members. I illustrated in Chapter 7 that positioning within these mining conflicts affected what responses to mining people saw as contributing to peace, as did more general notions of what could nourish community harmony. Although it is not possible to say that all Nasa miners saw mining as a requisite of peace, or that all opponents saw mining and peace as mutually exclusive, the miners felt aggrieved and misunderstood because their livelihoods depended on mining. Considering that some miners were ex-combatants and some others had political and economic grievances about the allocation of resources within their communities, responses to Nasa miners remained a contentious element with ongoing potential for escalating, violent conflict. As a result, there was a desire to see more dialogue and analysis to arrive at different solutions to the tensions instead of adding to existing territorial conflicts and to the violence of the post-agreement transition.

As I substantiated in Chapter 8, outside the Nasa community there was much concern about the nexus between mining, armed conflict and violence; about mining's human rights consequences; and about poor Government understanding of, and institutional capability to govern, mining. Like the Nasa, members of other rural communities saw strong links between uncontrolled mining, violence and the armed conflict. They expected to be consulted about new development projects and wanted to see different views of development coexist in their territories, but feared that instead the post-agreement period would see the arrival of mining entrepreneurs and intensify Government repression of dissenting communities. Specialists from the academic, human rights, multilateral, international aid and mining industry sectors shared concerns about Government's weak capacities, and lack of presence and political will to address the problems with mining. They were concerned that the Government failed to understand mining as a rights matter and to guarantee citizen and collective rights and participation, or to ensure a fair share of the benefits for the Nation and the host regions.

However, contrasting views signalled potential for intensifying conflicts about mining in the Colombian post-agreement context. Sectors in Government and industry were very optimistic in their assessment of Government capacity and industry potential, and shared a sense of urgency that legal barriers to the expansion of mining be removed. Mining, to members of those sectors, should be a force for the formalisation of rural Colombia and a source of funds to finance the post-agreement transition. In this scenario of community resistance to mining, limited Government capacity and understanding of the mining-violence-armed conflict links, and fiscal deficit and ambitions of foreign direct investment growth, new socio-environmental conflicts continued to emerge.

In looking at the Nasa case, with its multiple levels of specificity, I have examined how Nasa moral principles and Nasa interpretation of the Living Well discourse address questions of post-conflict mineral resource governance and peace, and how they shape community attitudes and responses to mining. I have illustrated that most Nasa disregard discourses about mining as a force for development because they have experienced its links with the violence and dispossession that hinder their ideals. I have elucidated that the Nasa active resistance identity in response to the armed conflict permeates their responses to mining. Nevertheless, rather than treating the Nasa response to mining as one of uniform resistance, I have identified internal contentions, tensions and problems that arise when Nasa people become miners, not least the potential to fuel internal conflict rather than build peace through prohibitive responses to mining. This has been possible because the aim was to understand the complexity of community views and responses rather than seek to classify them into pre-determined models of resource governance. The use of prescriptive models would have led to entirely different questions such as: How should the Nasa be consulted about mining? How can they better benefit from mining? Such questions can be relevant not as external impositions but within the scope of community understandings, experiences and readiness or willingness to address them.

The study demonstrates the relevance of moral precepts and interpretations in understanding Indigenous community participation in post-conflict mineral resource governance. Moral values and ideals shape people's expectations of the economy and community aspirations for the future and for peace. In the post-conflict environment of North Cauca Nasa territories, Nasa moral values embed knowledge about what the Nasa have endured and learned living amid armed conflict. Given the economic nature of the Colombian conflict, that type of knowledge is necessary to design appropriate approaches to manage natural resources while contributing to peace.

National and international actors and scholars interested in post-conflict mineral resource governance and peacebuilding need to give the study of moral understandings of mining a central position in their analysis. The invitation is to examine the moral values that sit behind discourses that advocate mining, in the abstract, as a positive force for development. In the governance prescription of strengthening the state and the rule of law, of getting the right mining operators, of distributing and utilising mining revenues well, there is embedded a simple assumption that mining is convenient per se. This assumption results from a set of specific moral principles. Proponents of those moral principles need to recognise how they shape their views on resource extraction. This shall prepare them to participate in the dialogues about morality that communities expected to host mining propose. This study encourages open discussion about the moral values behind opposition to, and advocacy for, mining.

Through this research I sought to bring attention to the active roles of local communities, such as the Nasa, in responding to mining dynamics in post-conflict environments, because only by understanding that active role can national and international actors contribute to positive mineral resource governance and peacebuilding outcomes. I have illustrated how local people cannot be left alone to respond to mining, in particular when it involves links to violent and criminal elements of society and when it builds on existing infrastructures of violence such as the ones the drug trade and the armed conflict provide in Colombia. However, it is also clear that external governance prescriptions cannot override Indigenous communities either. In the specific context of Colombia, this would carry serious consequences for peace and for collective rights. So, the central questions relate to where and how governments, but also other actors, take full responsibility, where they support or collaborate with community efforts, and where they should refrain from interfering or pushing boilerplate solutions. The answers to these questions are specific to the capacities and initiatives of each Indigenous community and that is where this research offers valuable insights.

In the case of the Nasa of North Cauca, National Government mineral resource governance efforts had been ill-suited to the mining reality the communities faced. Where Nasa communities decided to practice mining, they met blunt regulatory instruments and prohibitive costs for a community enterprise. The lack of finance and technical assistance made several Nasa mining ventures unviable. But where illegal miners entered Nasa territories, the State apparatus was often unable to exert control, leaving the community to assume the task and the cost – including the loss of life – that came with opposing illegal mining operations with criminal links. The Government, and often the broader State, failed to show up where required to halt the advance of illegal mining. In contrast, when communities were open to mining, there was no support available for them to experiment with the idea that mining could lead to community development. Echoing the statements of several participants from outside the Nasa community, it should not come as a surprise that most Nasa Authorities opposed mining and were concerned about the paths it could follow in post-agreement Colombia.

Future research on the moral understandings of, and the practical responses to, mining of other post-conflict communities would further increase knowledge on the complexities and requirements of post-conflict mineral resource governance and peacebuilding in Colombia. My hope is that research of this nature invites a more open debate on the moral principles that guide proponents and opponents of mining and inform national and international efforts on post-conflict mineral resource governance and peacebuilding. I also hope that increased attention to the specific community responses to mining in post-conflict environments might assist more collaborative approaches between local, national and international actors. This way, local communities can access and tailor international knowledge, rather than be expected to fulfil a prescribed role; and local experiences and knowledge can enrich global understandings and practices rather than be overlooked.

Index

Bold page numbers indicate tables.

For Product Safety Concerns and Information please contact our EU representative GPSR@taylorandfrancis.com
Taylor & Francis Verlag GmbH, Kaufingerstraße 24, 80331 München, Germany

www.ingramcontent.com/pod-product-compliance
Ingram Content Group UK Ltd.
Pitfield, Milton Keynes, MK11 3LW, UK
UKHW021307260926
475702UK00048B/854
* 9 7 8 1 0 3 2 1 2 9 2 6 6 *